Also by John Horgan

*The End of Science: Facing the Limits of Knowledge
in the Twilight of the Scientific Age*

THE
UNDISCOVERED

How the Human
Brain Defies
Replication,
Medication,
and Explanation

MIND

JOHN
HORGAN

A Touchstone Book
Published by Simon & Schuster
New York   London   Toronto   Sydney   Singapore

TOUCHSTONE
Rockefeller Center
1230 Avenue of the Americas
New York, NY 10020

First Touchstone Edition 2000
TOUCHSTONE and colophon are registered trademarks
of Simon & Schuster, Inc.

Manufactured in the United States of America
10   9   8   7   6   5   4   3   2   1

The Library of Congress has cataloged the Free Press edition as follows:
Horgan, John
    The undiscovered mind : how the human brain defies replication,
    medication, and explanation / John Horgan.
        p.   cm.
    Includes bibliographical references and index.
    1. Neurosciences Popular works.   I. Title.
    RC343.H636   1999   99-31051
    612.8'2—dc21   CIP
        ISBN 0-684-85075-3
            0-684-86578-5 (Pbk)

For my father

CONTENTS

I-WITNESSING

· ·

The sciences have developed in an order the reverse of what might have been expected. What was most remote from ourselves was first brought under the domain of law, and then, gradually, what was nearer: first the heavens, next the earth, then animal and vegetable life, then the human body, and last of all (as yet very imperfectly) the human mind.

—BERTRAND RUSSELL

Over the past few decades, anthropologists have cultivated a style of reportage in which they lay bare cultural, intellectual, or emotional predilections that might distort their observations. Clifford Geertz of the Institute for Advanced Study, who helped initiate this trend, has dubbed it "I-witnessing." By flaunting their subjectivity and thereby acknowledging that any pretense of objectivity is naive, if not deceitful, I-witnesses hope to earn more of the reader's trust. The effect, too often, is not only to distract the reader—who presumably is more interested in the sex lives of Fijians than in the homesickness of a Harvard doctoral candidate— but also to arouse rather than allay suspicions concerning the narrator's agenda. Confession is as dangerous a tactic for scientists as it is for politicians or lovers.

Or science writers. That said, I feel obliged to begin this book with a bit of I-witnessing because its subject—more than, say, particle physics or chaos theory—raises questions about the limits of objectivity. In the early 1990s, toward the end of my first decade as a science writer, I became increasingly disturbed by the way science was being portrayed by most scientists and science writers, myself included. For understandable reasons, researchers and reporters alike focus on the scientific frontiers that are generating the most advances, whether genuine or hypothetical.

This emphasis makes science seem more potent and fast moving than it really is. Paradoxically, the focus on frontiers also lends support to the postmodern proposition that science cannot deliver absolute, permanent truths, because all theories are provisional, subject to change. Both views overlook all the areas of science in which little or no progress is made, either because the major problems have already been solved and no fundamental mystery remains or because the problems have resisted all assaults. In my articles for *Scientific American*, where I was then employed, I began to shift my attention from science's accomplishments to its limitations.

My obsession with science's limits culminated in *The End of Science*, which was published in 1996. In it I examined major fields of pure science, including particle physics, cosmology, and evolutionary biology. These disciplines, I argued, were becoming victims of their own phenomenal success. Physicists would never transcend the powerful theories of quantum mechanics and relativity, which together describe all the forces and particles of nature; cosmologists would never achieve anything as profound as the unifying narrative of the big bang theory; biologists could not hope to top Darwin's theory of evolution and DNA-mediated genetics. But in the chapters titled "The End of Social Science" and "The End of Neuroscience," I presented a somewhat different argument: that scientists attempting to explain the human mind might be overwhelmed by its sheer complexity.

Here, as elsewhere, I took my cue from Gunther Stent of the University of California at Berkeley. In his prescient books *The Coming of the Golden Age* and *Paradoxes of Progress*, Stent argued that science was doomed to become a victim of both its success and its limitations. Stent recognized the importance of neuroscience, so much so that he switched to that field from molecular biology in the 1970s and he later served as the head of the neurobiology department of the National Academy of Sciences. But Stent was pessimistic about how far neuroscience could go toward explaining consciousness and other mysteries of the mind. He suspected that "the brain may not be capable, in the last analysis, of providing an explanation of itself."

Some critics of *The End of Science*, while granting that particle physics, cosmology, and evolutionary biology might be past their peaks, found my analysis of mind-related science unpersuasive, to put it politely. Lewis Wolpert, a pillar of British biology, was particularly offended by my handling of neuroscience. When I was introduced to him at a scientific gathering in London in 1997, Wolpert became so apoplectic that for a thrilling instant I thought he was going to strike me. The chapter on neuroscience in *The End of Science*, he shouted, his face reddening, was "appalling! Absolutely appalling!" It focused for the most part not on real neuroscientists but on latecomers like Gerald Edelman, an immunologist, and Francis Crick, who was originally a physicist! And how could I possibly say that neuroscience was ending when it was obviously just beginning!

Wolpert stalked off before I could respond. If our conversation had continued, I would have tried to appease him by granting that his objection to my book was not entirely unjustified. In fact, I had already decided that my treatment of mind-related science was insufficient, given the subject's enormous breadth and importance. *The End of Science* focused primarily on attempts to explain that most inescapable and elusive of all mental phenomena, consciousness. Consciousness is arguably the most philosophically resonant problem posed by the mind, but it is also arguably the most intractable and impractical problem.

Most investigators of the mind are pursuing other, more tangible questions: What processes in the brain allow us to see, hear, learn, remember, reason, emote, decide, act? Why are so many of us afflicted with mental disorders such as depression and schizophrenia? How effective are the medications, psychotherapies, and other remedies used to treat these devastating ailments? How do nature and nurture interact to produce an individual personality? What role did natural selection play in shaping our brains and minds? To what degree are we constrained by our biological heritage? Can the functions of the human mind be replicated by computers?

My argument that science has passed its peak was also based on a definition of science that implicitly slights mind-related fields and glorifies physics and cosmology. According to this definition, which I borrowed from particle physicists Steven Weinberg and Murray Gell-Mann, scientific truths can be ranked according to how broadly they apply through space and time. Quantum mechanics and general relativity are the most fundamental theories because they are true, as far as we know, throughout the entire universe. Within biology, Darwin's theory of evolution and DNA-based genetics are the most fundamental truths, because they apply—again, as far as we know—to all organisms that have ever lived on earth. In contrast, fields such as psychology, psychiatry, and behavioral genetics address just a single organism that has existed only for a few hundred thousand years or so.

On the other hand, these fields, which attempt to help us understand our own minds and behavior, are much more *meaningful* to most of us than physics or cosmology. As the physician Sherwin Nuland wrote in his meditation on mortality, *How We Die,* "I am more concerned with the microcosm than the macrocosm. I am more interested in how a man lives than how a star dies; how a woman makes her way in the world than how a comet streaks across the heavens. . . . The human mystery engages my fascination, not the condition of the cosmos." Narcissists that we are, there is no subject that fascinates us more than ourselves.

Mind-related science is not *merely* meaningful, however. In a strictly practical sense, the study of *Homo sapiens* is the most momentous of all scientific enterprises. Even pseudoscientific proclamations concerning human nature have the power to alter the course of history. The movements founded by Karl Marx and

Sigmund Freud—and by Jesus, Buddha, and Muhammad, for that matter, whose theologies also incorporated implicit theories of human nature—have demonstrated as much.

War, poverty, pollution, crime, racism, and indeed almost all our social woes, originate at least in part within our brains. So do depression, panic disorder, schizophrenia, and alcoholism. According to the World Health Organization, more than 1.2 billion people suffer from some type of neuropsychiatric or behavioral ailment. The annual costs of brain-related disorders in the United States alone exceed $300 billion, more than the estimated costs of cancer, heart disease, and AIDS combined.

The solutions to these problems may emerge from human minds as well. If neuroscientists, psychologists, artificial intelligence researchers, and other investigators of the psyche realize all their dreams, we may one day live in a culture shaped by *true* theories of human nature. We may no longer fret over the nature-nurture conundrum or the mind-body problem, because they will have been resolved to everyone's satisfaction. We may know enough about our own natures to design a political system that minimizes misery and maximizes happiness. We may have at our disposal drugs that dissipate despair and amplify memory, genetic therapies that abolish manic depression and boost intelligence. We may be served by robots as clever and charming as *Star Trek*'s Commander Data. We may *become* robots as clever and charming as Commander Data.

For all these reasons I decided to write another book, one that would examine mind-related science in much greater detail than *The End of Science* did. The book would address not only scientists' efforts to explain the properties of the mind, including consciousness; it would also examine attempts to medicate or otherwise treat minds afflicted with mental illness and to replicate the mind's properties in machines.

Science Versus Mind-Science

In *The End of Science* I coined the term *ironic science* to describe science that never gets a firm grip on reality and thus does not converge on the truth. Ironic science does not make the kind of literal, factual statements about the world that can be either confirmed or invalidated through empirical means; it is thus more akin to philosophy or literary criticism or even literature than to true science. Ironic science crops up in the so-called hard sciences, such as physics, astronomy, and chemistry. (An obvious example of ironic science is a theory that postulates the existence of other universes in addition to our own.) But ironic science is most pervasive in those fields that address the human mind.

Mind-related science—a graceless term that I will occasionally contract to *mind-science*—poses a special challenge to researchers seeking tangible, durable truths. The evolutionary biologist Ernst Mayr of Harvard University has pointed out that no field of biology can match the precision and power of physics, because unlike electrons or neutrons, all organisms are unique. But the differences between, say, two *Escherichia coli* bacteria or two leafcutter ants are trivial compared to the differences between any two humans, even those who are genetically identical. Each individual mind may also change dramatically when its owner is spanked, learns the alphabet, reads *Thus Spoke Zarathustra*, takes LSD, falls in love, gets divorced, undergoes Jungian dream therapy, suffers a stroke. The variability and malleability of minds enormously complicate the search for general principles of human nature.

Investigations of the mind have also failed to generate the kinds of applications that compel belief in a particular paradigm. Physicists can boast about lasers, transistors, radar, jets, nuclear bombs. Biologists can show off vaccines, antibiotics, cloning, and other marvels. The by-products of mind-science are rather less impressive: cognitive behavioral therapy, Thorazine, Prozac, shock therapy, alleged genetic markers for homosexuality, IQ tests, chess-playing computers.

The philosopher Thomas Kuhn contended that modern scientific theories are not *truer* than the theories they displace but merely *different*. Kuhn's proposal simply does not apply to certain fields of science, such as astronomy. In the late nineteenth century, astronomers believed that the smudges of light in the sky known as nebulas were clouds of gas in our own galaxy, the Milky Way. As telescopes became more powerful, astronomers realized that each nebula is an entire galaxy in its own right lying far beyond the Milky Way's borders. That is not just a different view; it is the *right* view.

But Kuhn's model of scientific nonprogress applies rather well to mind-science. Clifford Geertz, the same anthropologist who coined the term *I-witnessing*, mused recently that psychology has "been driven in wildly different directions by wildly different notions of what it is, as we say, 'about'—what sort of knowledge, of what sort of reality, to what sort of end, it is supposed to produce. . . . Paradigms, wholly new ways of going about things, come along not by the century, but by the decade; sometimes, it almost seems, by the month."

Theories of human nature never really die; they just go in and out of fashion. Often old ideas are simply repackaged in more palatable forms. Phrenology is reincarnated as cognitive modularism. Sociobiology mutates into evolutionary psychology. Eugenics, stripped (for the most part) of its unsavory political tenets, evolves into behavioral genetics. Old treatments linger too. Shock treatments and lobotomies, pushed to the sidelines of psychiatry in recent decades by Prozac and lithium, are still prescribed for severe mental illness.

One paradigm that has demonstrated a Rasputin-like persistence is psychoanalysis, which Freud invented a century ago. Although psychoanalysis has declined in prestige over the past few decades, millions of people still receive psychotherapy based—at least indirectly—on Freudian tenets. Moreover, many intellectuals—including not only French philosophers but also neuroscientists, artificial intelligence researchers, and others who supposedly should know better—still profess admiration for psychoanalysis. Why does psychoanalysis, once defined as the "treatment of the id by the odd," remain so influential? In response to this question, Freudophobes denounce Freud as a cult leader who excelled at nothing so much as self-promotion; Freudophiles hail him as a genius whose insights into the psyche, though difficult to pin down empirically, still ring true. Each of these views is defensible, but each also overlooks the crucial factor underlying the persistence of psychoanalysis: the inability of science to offer an obviously superior explanation of the mind and its disorders. Freudians cannot point to unambiguous evidence of their paradigm's superiority, but neither can proponents of more modern paradigms.

The anti-Freudians argue, in effect, that psychoanalysis has no more scientific standing than phlogiston, the pseudo-substance that eighteenth-century physicists believed was released during combustion. But the reason physicists do not still debate the phlogiston hypothesis is that the discovery of oxygen and other advances in chemistry and thermodynamics rendered it utterly obsolete. A century's worth of research in psychiatry, genetics, neuroscience, and adjacent fields has not yielded a paradigm powerful enough to obviate Freud once and for all. If psychoanalysis is the equivalent of phlogiston, as the anti-Freudians claim, so are all its would-be successors. As Thomas Kuhn might have put it, these alternatives are not truer or better; they're just different.

The adversary system is essential to science, as it is to law. But the fractiousness of mind-science sets it apart from other fields. Researchers are often most earnest and enthusiastic not when touting their own pet paradigm but when bashing the paradigms of others. Thus, a neuroscientist mocks evolutionary psychology as a grab-bag of just-so stories. A proponent of electroshock therapy dwells on the sexual side effects of Prozac. A behavioral geneticist derides the robotic fantasies of artificial intelligence researchers. Even among researchers committed to a single paradigm, the cross-fire can be lethal.

I have exploited this internecine conflict for my own purposes. Some readers may find it unfair and inconsistent for me to criticize, say, behavioral genetics in one chapter and then invoke it in the next chapter to raise doubts about evolutionary psychology. But when it comes to theories of human nature, standards of proof should be the inverse of those that apply in courts. Theories should be considered guilty—that is, wrong or dubious—until their correctness is established beyond a reasonable doubt. Critics should therefore be allowed to cast doubt on a theory by introducing a contradictory hypothesis that may be equally dubious. Moreover, the fact that competent scientists adhere to such different, contradictory paradigms is grounds for skepticism toward all the paradigms.

Scientists often defend neuroscience and related fields by claiming that they are "just beginning," as my British antagonist Lewis Wolpert put it. Actually neuroscience has a history comparable to that of any other field of science. In the fifth century B.C., Hippocrates hypothesized that the brain is the seat of human perception and thought, and Galen confirmed this supposition roughly six centuries later. Luigi Galvani showed in the late eighteenth century that nerves emit and respond to electric current, and around the same time Franz Joseph Gall invented phrenology, the ancestor of the modular-mind theory that cognitive scientists and others are now touting. Francis Galton sought to solve the nature-nurture debate by studying identical twins in the mid–nineteenth century. William James wrote *Principles of Psychology* in 1890, and Freud began setting forth his psychoanalytic theory shortly after—having already written a solid monograph on language disorders caused by brain damage. Meanwhile, Camillo Golgi, Santiago Ramon y Cajal, and others were unraveling the structure and function of neurons.

The claim that neuroscience is "just beginning" is based not on the field's actual age but on its *productivity*. Wolpert acknowledged as much in his 1993 book, *The Unnatural Nature of Science*. Mind-related research is still in a "primitive" state, Wolpert wrote, compared to more mature scientific fields such as nuclear physics and molecular biology. As evidence, Wolpert cited the inability of neuroscientists to either confirm or falsify the propositions of psychoanalysis: "It is . . . not possible, at the present time, to do any experiment at a lower level of organization—that is, at the level of brain function or neurophysiology—which would contradict psychoanalytic theory." Yet Wolpert firmly rejected the pessimistic view that "human behavior and thought will never yield to the sort of explanations that are so successful in the physical and biological sciences." He argued that "we just do not know what we do not know and hence what the future will bring."

I agree with Wolpert on the current "primitive" state of neuroscience and other mind-related fields. The question is, Just how far will mind-science go in the future, given how little progress there has been to date? Like most other scientists, Wolpert is an optimist. He contends essentially that the *lack* of progress in mind-science thus far means that great things await us. In other words, past failure predicts future success. This is not so much an argument as an expression of faith. Given their poor record to date, I fear that neuroscience, psychology, psychiatry, and other fields addressing the mind might be bumping up against fundamental limits of science. Scientists may never completely succeed in healing, replicating, or explaining the human mind. Our minds may always remain, to some extent, undiscovered.

What's the Upside?

Before I started writing this book, I bounced the idea off a literary agent who specializes in science. "Yup, I got it," he said after I had rattled off all the fields I intended to critique. "Okay, so you give us all this negative stuff. Now, what's going to be the upside?" Upside? I asked. "Yeah, you know, the positive message. What are you going to give people at the end of your book so they don't go away depressed?" The question caught me off guard. "I don't do upside," I said, before muttering something about the truth being its own reward. Although the agent looked dubious, he didn't press the issue.

As I began working on the book, I didn't become any less pessimistic about the current state of mind-science. Quite the contrary. But I did feel more of a need to justify my critical stance, to find an upside. Early in my research, I observed patients at the New York State Psychiatric Institute receiving shock therapy. One of the patients was a slim, delicate-looking woman with short brown hair. She lay on a gurney awaiting her treatment, while I stood a few feet away scribbling on a pad of yellow paper. As a technician rubbed conductive jelly on the woman's temples, she suddenly turned her head and stared directly at me. She seemed simultaneously puzzled, frightened, and angry, as if she were thinking, *Who the hell are you, and what are you doing here watching me suffer?*

I felt a similar pang of guilt when I ran into a childhood acquaintance, whom I'll call Harry, at a party. We hadn't seen each other in years. He asked what I was up to, and I told him about my book project, emphasizing, as I often did with nonscientists, my planned critique of medications like Prozac. As I spoke, Harry looked increasingly uncomfortable and finally told me why. A few years earlier, he had fallen into a depression so deep that he had considered suicide. Prozac had pulled him out of it. Without it, he might be dead. What did I hope to accomplish, he gently asked me, by denigrating the drug that had saved him and many others?

Another issue arose when I previewed some of the themes of this book in a lecture at a university in California. During the question-and-answer period a geneticist angrily asked me what my point was. Did I think he and his colleagues should simply give up? Should Congress take their funding away? These encounters nudged my attention away from my subject matter and toward my own attitude. Why was I so negative? What was my motive? Did I actually *want* this research to fail? Assuming that my view of mind-science was correct, what did I hope to accomplish by stating it? What good would it do? As that literary agent had put it, *What's the upside?*

I'll take up the concerns of my childhood acquaintance Harry first. It concerns me that criticism of Prozac, psychotherapy, and other remedies might undermine belief in them and thus diminish their effectiveness for people like Harry; science, after all, has shown that faith in a given therapy can become self-fulfilling. But surely it would be irresponsible, even cruel, for a journalist knowingly to exaggerate the effectiveness of a treatment so that someone might reap more benefit from it. According to this reasoning, journalists might as well tout the healing powers of leeches, crystals, or homeopathy. Faith is not omnipotent, nor is it always benign. Religious faith is arguably the most successful psychological therapy ever invented, but it has also fomented ignorance and intolerance. The benefits of scientific knowledge *must* outweigh the benefits of faith. Otherwise, why practice science at all?

Proclamations about the limits of neuroscience, behavioral genetics, and related fields could also conceivably discourage scientists from pursuing this type of research and dissuade public officials from allocating funds for further inquiry. But again, that possibility cannot justify ignoring or misrepresenting the facts. My intention in writing this book is to provide constructive criticism of mind-science, which is potentially the most important of all scientific endeavors. It is precisely because this research is so important that it demands scrutiny. My criticism might seem harsh at times, but that is because I wish to redress an imbalance; most books about mind-science, whether by researchers or journalists, are written in a celebratory rather than a critical mode.

Some problems addressed by mind-science may be intractable, but I would hate to see this prognosis become self-fulfilling. In spite of their missteps and limitations, neuroscience, psychology, psychiatry, behavioral genetics, evolutionary psychology, artificial intelligence, and, yes, even psychoanalysis are far from worthless. Each has yielded clues to our nature, albeit ambiguous and even contradictory ones. If nothing else, each of these fields can serve as a counterweight to the others, ensuring that none becomes too powerful. Someday, moreover, scientists might actually begin to understand human nature and find ways to improve it.

But the faith and optimism that scientists need to sustain such a difficult quest can also get them into trouble. In the past, excessive belief in the power of science and reason led to pseudo-scientific ideologies such as social Darwinism, eugenics, and totalitarian communism. I would like to believe that scientists—and the rest of us—have learned by now not to give too much credence to any particular theory, but I see too many signs to the contrary. I am disturbed by the proliferation of recovered-memory therapy; the increasingly widespread treatment of children with psychiatric drugs; the persistence of racist theories of intelligence; the promulgation of cartoonish depictions of male and female sexuality. More subtle harm can come from the suggestions of prominent researchers that we humans are *just* a pack of neurons or *just* vehicles for propagating genes or *just* machines. This kind of reductionism does a disservice to both humans and science.

When it comes to human nature, our lust for absolute truths, unified theories, and panaceas can have dangerous consequences. The trick is to be skeptical of science's products while remaining supportive of the scientific enterprise. The philosopher Karl Popper embodied this attitude. We can never prove our theories are true, Popper proposed; we can only disprove theories, or falsify them. All knowledge is thus tentative, provisional—and science, conveniently for the science-loving Popper, becomes an immortal enterprise. In *The End of Science*, I argued that Popper's scheme cannot be sustained when applied to all of science. Much of the knowledge we have gleaned from physics, astronomy, and biology is not provisional but permanent and absolute, as much so as the fact that the earth is round and not flat. But when applied to mind-science, which has a relatively weak grip on reality, Popper's philosophy makes a great deal of sense.

Popper called his philosophy *critical rationalism*. I prefer the term *hopeful skepticism*. Too little skepticism leaves us prey to peddlers of scientific snake oil. Too much skepticism can lead to solipsism, to a radical postmodernism that denies the possibility of achieving not only complete knowledge of our selves but any knowledge at all. But just the right amount of skepticism—mixed with just the right amount of hope—can protect us from own lust for answers while keeping us open-minded enough to recognize genuine truth if and when it arrives. If this book succeeds at all, it will persuade readers to view mind-science with hopeful skepticism. That's one upside.

Even if the mind continues to defy the efforts of scientists to explain, heal, and replicate it—even if it remains undiscovered—there is another upside. Science has provided humanity with a grand, ennobling purpose; if this quest for knowledge ceases, we will lose something precious. The goals of mind-science are so alluring that researchers will surely never stop pursuing them, nor will governments, companies, and philanthropies ever stop funding the pursuit. The fact that these goals may never be fully attained means, paradoxically, that mind-related science might continue forever. As long as we remain mysteries to ourselves, as long as we suffer, as long as we have not descended into a utopian torpor, we will continue to ponder and probe our minds with the instruments of science. How can we not? Inner space may be science's final—and eternal—frontier.

1

NEUROSCIENCE'S EXPLANATORY GAP

. . . .

By 1979 Freudian psychology was treated as only an inter-esting historical note. The fashionable new frontier was the clinical study of the central nervous system. . . . Today the new savants probe and probe and slice and slice and project their slides and regard Freud's mental constructs, his "li-bidos," "Oedipal complexes," and the rest, as quaint quack-eries of yore, along the lines of Mesmer's "animal magnetism."

—TOM WOLFE, *IN OUR TIME*

In *Phaedo* Plato described the last hours of Socrates, who had been imprisoned and sentenced to death by Athenian authorities. Socrates told friends who had assembled in the prison why he had accepted his death sentence rather than fleeing. At one point, Socrates ridiculed the notion that his behavior could be explained in physical terms. Someone who held such a belief, Socrates speculated, would claim that

> as the bones are lifted at their joints by the contraction or re-laxation of the muscles, I am able to bend my limbs, and that is why I am sitting here in a curved posture . . . and he would have a similar explanation of my talking to you, which he would at-tribute to sound, and air, and hearing . . . forgetting to mention the true cause, which is, that the Athenians have thought fit to

condemn me, and accordingly I have thought it better and more right to remain here and undergo my sentence.

This is the oldest allusion I know of to what some modern philosophers call the explanatory gap. The term was coined by Joseph Levine, a philosopher at North Carolina State University. In "Materialism and Qualia: The Explanatory Gap," published in *Pacific Philosophical Quarterly* in 1983, Levine addressed the puzzling inability of physiological theories to account for psychological phenomena. Levine's main focus was on consciousness, or "qualia," our subjective sensations of the world. But the explanatory gap could also refer to mental functions such as perception, memory, reasoning, and emotion—and to human behavior.

The field that seems most likely to close the explanatory gap is neuroscience, the study of the brain. When Plato wrote *Phaedo*, no one even knew that the brain is the seat of mental functioning. (Aristotle's observation that chickens often continue running after being decapitated led him to rule out the brain as the body's control center.) Today neuroscientists are probing the links between the brain and the mind with an ever more potent array of tools. They can watch the entire brain in action with positron emission tomography and magnetic resonance imaging. They can monitor the minute electrical impulses passing between individual nerve cells with microelectrodes. They can trace the effects of specific genes and neurotransmitters on the brain's functioning. Investigators hope that eventually neuroscience will do for mind-science what molecular biology did for evolutionary biology, placing it on a firm empirical foundation that leads to powerful new insights and applications.

Neuroscience is certainly a growth industry. Membership in the Society for Neuroscience, based in Washington, D.C., soared from 500 in 1970, the year it was founded, to over 25,000 in 1998. Neuroscience journals have proliferated, as has coverage of the topic in premier general-interest journals such as *Science* and *Nature*. When *Nature* launched a new periodical, *Nature Neuroscience*, in 1998, it proclaimed that neuroscience "is one of the most vigorous and fast growing areas of biology. Not only is understanding the brain one of the great scientific challenges of our time, it also has profound implications for society, ranging from the basis of memory to the causes of Alzheimer's disease to the origins of emotions, personality and even consciousness itself." Neuroscience is clearly advancing; it is getting somewhere. But where?

I once asked Gerald Fischbach, the head of Harvard's Department of Neuroscience and a former president of the Society for Neuroscience, to name what he considered to be the most important accomplishment of his field. He smiled at the naiveté of the question. Neuroscience is a vast enterprise, he pointed out, which ranges from studies of molecules that facilitate neural transmission to magnetic resonance imaging of whole-brain activity. It is impossible, Fischbach added, to single out any particular finding, or even a set of findings, emerging from neuroscience. The field's most striking characteristic is its production of such an enormous and still-growing number of discoveries. Researchers keep finding new types of brain cells, or neurons; neurotransmitters, the chemicals by which neurons communicate with each other; neural receptors, the lumps of protein on the surface of neurons into which neurotransmitters fit; and neurotrophic factors, chemicals that guide the growth of the brain from the embryonic stage into adulthood.

Not long ago, Fischbach elaborated, researchers believed there was only one receptor for the neurotransmitter acetylcholine, which controls muscle functioning; now at least ten different receptors have been identified. Experiments have turned up at least fifteen receptors for the so-called GABA (gamma-amino butyric acid) neurotransmitter, which inhibits neural activity. Research into neurotrophic factors is also "exploding," Fischbach said. Researchers had learned that neurotrophic factors continue to shape the brain not only in utero and during infancy but throughout our life span. Unfortunately, neuroscientists had not determined how to fit all these findings into a coherent framework. "We're not close to having a unified view of human mental life," Fischbach said.

Fischbach was spotlighting one of his field's most paradoxical features. Although *reductionist* is often used as a derogatory term, science is reductionist by *definition*. As the philosopher Daniel Dennett once put it, "Leaving something out is not a feature of failed explanations, but of successful explanations." Science at its best isolates a common element underlying many seemingly disparate phenomena. Newton discovered that the tendency of objects to fall to the ground, the swelling and ebbing of seas, and the motion of the moon and planets through space could all be explained by a single force, gravity. Modern physicists have demonstrated that all matter consists basically of two types of particles, quarks and electrons. Darwin showed that all the diverse species on earth were created through a single process, evolution. In the last half-century, Francis Crick, James Watson, and other molecular biologists revealed that all organisms share essentially the same DNA-based method of transmitting genetic information to their offspring. Neuroscientists, in contrast, have yet to achieve their reductionist epiphany. Instead of finding a great unifying insight, they just keep uncovering more and more complexity. Neuroscience's progress is really a kind of anti-progress. As researchers learn more about the brain, it becomes increasingly difficult to imagine how all the disparate data can be organized into a cohesive, coherent whole.

The Humpty Dumpty Dilemma

In 1990, the Society for Neuroscience persuaded the U.S. Congress to designate the 1990s the Decade of the Brain. The goal of the proclamation was both to celebrate the achievements of neuroscience and to support efforts to understand mental disorders such as schizophrenia and manic depression (also known as bipolar illness). One neuroscientist who opposed the idea was Torsten Wiesel, who won a Nobel prize in 1981 and went on to become president of Rockefeller University in New York. (He stepped down to return to research at the end of 1998.) Born and raised in

Sweden, Wiesel is a soft-spoken, reticent man, but when I interviewed him at Rockefeller University in early 1998, he became heated at the mention of the "Decade of the Brain."

The idea was "foolish," he grumbled. "We need at least a century, maybe even a millennium," to comprehend the brain. "We still don't understand how *C. elegans* works," he continued, referring to a tiny worm that serves as a laboratory for molecular and cellular biologists. Scientists had discovered some "simple mechanisms" in the brain, but they still did not really understand how the brain develops in the womb and beyond, how the brain ages, how memory works. "We are at the very early stage of brain science." (Nevertheless, in 1998 behavioral scientists—a broad category including psychologists, geneticists, anthropologists, and others—began lobbying for the decade beginning in the year 2000 to be named the Decade of Behavior.)

Wiesel himself participated in one of neuroscience's paradigmatic discoveries. Like many other scientific triumphs, this one resulted from both hard work and serendipity. In 1958 Wiesel and another young neuroscientist, David Hubel, were conducting experiments on the visual cortex of a cat in a "small, dingy, windowless basement lab" (according to one account) at the Johns Hopkins Medical School. After implanting an electrode in the cat's visual cortex, they projected images on the cat's retina with a slide projector attached to an ophthalmoscope. They presented the cat with two simple stimuli: a bright spot on a dark background and a dark spot on a bright background. When the electrode detected an electric discharge from a neuron, a device similar to a Geiger counter would emit a loud click.

Wiesel and Hubel were getting inconclusive results when one of their slides became stuck in the projector. After unjamming the slide, they slowly pushed it into its slot. Suddenly the electrode monitor started firing like "a machine gun." Wiesel and Hubel eventually realized that the neuron was responding to the edge of the slide moving across the cat's field of vision. In subsequent experiments, they found neurons that respond to lines only at specific orientations relative to the position of the retina. As the investigators moved the electrode through the visual cortex, the orientation of the lines to which the neurons responded kept changing, like a minute hand circling a clock. In 1981 Wiesel and Hubel received a Nobel prize for their research.

These findings are emblematic of a larger trend in neuroscience. Arguably the most important discovery to emerge from the field is that different regions of the brain are specialized for carrying out different functions. This insight is hardly new; Franz Gall said as much two centuries ago when he invented phrenology (which degenerated into a pseudoscientific method for determining character from the shape of the skull). But modern researchers keep slicing the brain into smaller and smaller pieces, with no end to the process in sight.

As recently as the 1950s, many scientists believed that memory is a single—albeit highly versatile—function. The researcher Karl Lashley was a prominent advocate of this view. He argued that memories are processed and stored not in any single location but throughout the brain. As evidence, he pointed to experiments in which lesions in the brains of rats did not significantly affect their ability to remember how to navigate mazes. What Lashley failed to realize was that rats have many redundant methods for navigating a maze; if the rat's ability to recollect visual cues is damaged, it may rely instead on olfactory or tactile cues.

Subsequent experiments involving both humans and other animals revealed many types of memory, each underpinned by its own region of the brain. The two major categories of memory are explicit, or declarative, memory, which involves conscious recollection; and implicit, or nonconscious, memory, which remains below the level of awareness but nonetheless affects behavior and mental functioning.

Memory has been divided into other categories as well, some of which overlap. Short-term memory, which is sometimes called working memory, allows us to glance at a telephone number and recall it just long enough to dial it a few seconds later. Long-term memory keeps that same telephone number in permanent storage, ready to be accessed when needed. Procedural memory lets us acquire and perform such reflexive skills as driving a car, touch-typing, or playing tennis. Episodic memory enables us to recall specific events.

Experiments have also identified a phenomenon known as priming, which is similar to the old notion of subliminal influence. Subjects are exposed to a stimulus, such as a sound or image, so briefly that they never become consciously aware of it and cannot recall it later. Yet tests show that the stimulus has been imprinted on the brain at some level. In one set of experiments, subjects are shown a list of words too briefly for it to be stored in short-term memory. Later the subjects are asked to play a game similar to the television game *Wheel of Fortune*. Given the clue "o-t-p-s," they must guess what the full word is. Subjects who have previously been exposed to a list of words containing *octopus* are much more likely to guess correctly, even though they cannot explicitly recall whether the list included *octopus*.

Technologies such as positron emission tomography (PET) and magnetic resonance imaging (MRI) have accelerated the fragmentation of the brain and mind. PET scans monitor short-lived radioactive isotopes of oxygen that have been injected into the blood. High levels of the isotope indicate increased blood flow and thus increased neural activity. MRI dispenses with the need for an injection of a radioactive substance. A powerful electromagnetic pulse causes certain atoms to align in a particular direction, like iron filings arranged around a magnet. When the magnetic field is relaxed, the atoms emit radiation at characteristic frequencies.

Imaging studies often focus on subjects performing some task: solving mathematical puzzles, sorting images according to category, memorizing lists of words. Those regions of the brain that are most active are assumed to be crucial to the activity. Karl Friston, an MRI specialist at the Institute of Neurology in London, compared this cataloguing of neural "hot spots" to Darwin's patient gathering of data on animals from around the world. "Without this catalogue of functional specialization," Friston said, "I don't think that one's going to go far in assembling a useful and accountable theory of brain organization."

But Friston felt that the push toward localization had gone too far. Too many studies simply associate a given region with a given function "without any reference to any conceptual framework or proper or deep understanding of the functional architecture of the brain." Different parts of the brain are also clearly interconnected, and understanding these neural connections is crucial to understanding the mind. "Looking at the correlations *between* different areas," Friston said, "has been very much underemphasized."

Rodolfo Llinas, a neuroscientist at New York University, was even more critical of the manner in which neuroimaging is being used, particularly in psychiatry. "You find somebody who has a particular problem, and you see a red spot on the front of the cortex and you say, 'Okay, so that spot of the cortex is the site where you have bad thoughts.' It's absolutely incredible! The brain does not function as a single-area organ!" Llinas compared these studies to phrenology, the eighteenth-century pseudoscience that divided the brain into discrete chunks dedicated to specific functions. "You have a patient, and you put the patient into the instrument, and you write a paper, because you can just see it," Llinas said. "It's phrenology!"

Llinas recalled that neuroscience previously went through a phase when researchers injected drugs into monkeys or rats and published a paper on the results, whether or not the results were meaningful. "We're about in those terms" with the new imaging technologies, Llinas asserted. "We tend to publish a few cases and to say, 'This is how it works, because look at the beautiful picture it got.'" But "then you go into the details, and it becomes a bit of a mirage."

As neuroscientists keep subdividing the brain, one question looms ever larger: How does the brain coordinate and integrate the workings of its highly specialized parts to create the apparent unity of perception and thought that constitutes the mind? The Harvard neuroscientist David Hubel, whose experiments with

Torsten Wiesel helped to create the current crisis in neuroscience, stated at the end of his book *Eye, Brain and Vision:*

> This surprising tendency for attributes such as form, color, and movement to be handled by separate structures in the brain immediately raises the question of how all the information is finally assembled, say for perceiving a bouncing red ball. It obviously must be assembled somewhere, if only at the motor nerves that subserve the action of catching. Where it's assembled, and how, we have no idea.

This conundrum is sometimes called the binding problem. I would like to propose another term: the Humpty Dumpty dilemma. It plagues not only neuroscience but also evolutionary psychology, cognitive science, artificial intelligence—and indeed all fields that divide the mind into a collection of relatively discrete "modules," "intelligences," "instincts," or "computational devices." Like a precocious eight-year-old tinkering with a radio, mind-scientists excel at taking the brain apart, but they have no idea how to put it back together again.

Patricia Goldman-Rakic's Explanatory Gap

One neuroscientist striving to solve the Humpty Dumpty dilemma is Patricia Goldman-Rakic, a professor at the Yale University School of Medicine. Goldman-Rakic, who heads one of the most sophisticated neuroscience laboratories in the world, studies not human brains but those of a close relative, the macaque monkey. Goldman-Rakic calls herself a "systems neuroscientist." By working on the frontal cortex, which is thought to be the seat of reasoning ability, decision making, and other higher cognitive functions, she hopes to show how psychology, psychiatry, and other macro-level approaches to the mind can be integrated with the more reductionist models focusing on neural, genetic, and molecular processes.

A major focus of her research is working memory. Like the random-access memory of a computer, which makes information available for instant use, working memory allows us to maintain the thread of a conversation, read a book, play a game of cards, or perform simple arithmetic calculations in our heads. Many neuroscientists think a better understanding of working memory will help to solve mysteries such as the binding problem, free will, consciousness, and schizophrenia. No other neuroscientist is better positioned to close the explanatory gap than Goldman-Rakic, and yet I never felt the explanatory gap more vividly, even viscerally, than when I visited her laboratory.

The animal rights movement has turned laboratories such as Goldman-Rakic's into fortresses. Visitors must check in with an armed security guard at the entrance of the Yale Medical School; they are escorted through two steel doors, each with a small window, which can be opened only with a magnetic key. Within is a large suite of rooms containing monkeys, microscopes, surgical equipment, and all the latest instruments of the biotechnology revolution. In one room, a young woman was painstakingly slicing the frozen, walnut-size brain of a monkey into transparently thin sheets with what looked like a miniature deli slicer. A young man in an adjacent office was examining cross-sections in a microscope and tracing on paper the fantastically intricate connections between the neurons. Later he would transfer these tracings into a computer to form high-resolution, three-dimensional maps of neural circuitry.

Goldman-Rakic and her colleagues have perfected a technique that provides the same information as a PET scan but with much higher resolution. After being injected with radioactive chemicals that help to metabolize glucose, the monkeys perform certain tasks. Investigators quickly sacrifice the monkeys and freeze their brains. By measuring the levels of radioactivity in different regions of the brain, the researchers can determine which regions contributed most to the performance of the task.

Another room houses an apparatus for probing the working memory of monkeys. The monkey sits on a chair in a box-shaped steel frame facing a screen on which the researchers project signals and images. Its head is fixed in place with bolts that are screwed into its skull and attached to the frame. A sensor implanted in the monkey's eye—the wire from which passes through a plug in the monkey's skull to a recording device—allows the researchers to track eye movements. Electrodes implanted in the monkey's frontal cortex monitor the firing of individual neurons.

The master of this rather forbidding domain is a petite, elegantly coiffed woman who, on the day of my visit, wore a white cashmere sweater and gold earrings. When we sat down to discuss her work, Goldman-Rakic was for the most part guarded and reserved, but now and then to emphasize a point she leaned toward me and gripped my forearm. The goal of her research, she said, is to understand such higher cortical functions as memory, perception, and decision making. For those interested in higher cortical functions, the macaque monkey serves as an "unexcelled" model, she said. Monkeys are capable of cognition that is fundamentally similar to that of humans, though obviously not as sophisticated. When injected with amphetamines, monkeys even display behavior that resembles that of schizophrenic humans. "We're at the edge," Goldman-Rakic said, "making discoveries that are of great moment for understanding humans."

Experiments on monkeys have helped to illuminate working memory, which Goldman-Rakic described as a "mental sketchpad" or "glue" that helps to provide continuity of thought. Working-memory capacity correlates strongly with general intelligence and reading ability. People with a weak working memory have a harder time understanding complex sentences, in which subjects and verbs are separated by embedded clauses. Schizophrenia may also stem from a deficit in working memory. One major symptom of schizophrenia is "thought derailment," Goldman-Rakic explained. Schizophrenics keep losing their train of thought; they are therefore excessively sensitive to and easily overwhelmed by incoming perceptions.

Her research could provide insights into both normal and deranged human cognition and thus point the way to better pharmacological or behavioral therapies. She and her coworkers were studying how dopamine, serotonin, and other neurotransmitters inhibit or facilitate cortical functioning. "Many diseases involve dopamine: schizophrenia, Parkinson's disease, possibly childhood disorders like attention deficit syndrome." Drugs such as Prozac suggested that serotonin can have a profound effect on mood. Prozac "changes a person's life from blackness to lightness, all right? Now, why?" Her team had just taken one step toward the answer by showing that certain cortical cells react differently to incoming signals depending on serotonin levels.

Could her research lead to drugs that boost memory and intelligence? "Absolutely! Definitely!" she exclaimed. "There are drugs that do this already, but these are drugs that are not necessarily always effective, or they have side effects." She emphasized that her group is not pursuing such drugs as an end in themselves. "The goal of my research is not to support the pharmaceutical industry, at all. It's to learn how the brain works, and particularly how the portions of the brain or the systems that are involved in cognition work."

Cognition, explained Goldman-Rakic, entails much more than merely responding automatically to a stimulus, like a driver stopping at a red light and going on green. "Humans have lots of habitual responses, automatic responses, reflexive responses. But that's not what makes them human. What makes them human is the *flexibility* of their responses, their ability *not* to respond as well as to respond, their ability to reflect, and their ability to draw upon their experience, to guide a particular response at a particular moment." Was she really talking about free will? "I could use that terminology," Goldman-Rakic replied, dropping her voice and speaking in a conspiratorial mock whisper, "if I really were disinhibited."

She fetched an article describing one of her experiments and opened it on the table in front of us. In the experiment, the monkey was trained to keep his eyes focused on the center of a screen while the researchers briefly shone a light on one of the screen's edges or corners. The monkey had learned to wait a few seconds after the light went off before looking directly at where the light had been. During these few seconds, the monkey had to store the light's location in his working memory.

Goldman-Rakic pointed to one of the article's graphs, which represented the activity of neurons that started firing when the light appeared and kept firing after the light vanished. She noted that Torsten Wiesel and David Hubel and most other neuroscientists focused on neurons that responded directly to external stimuli. "This," Goldman-Rakic said, jabbing her finger at the graph, "is sooooo different." These neurons were firing in the *absence* of an external stimulus; this neural activity corresponded not to a real image but to the memory, or internal representation, of an image. "This," she continued dramatically, "is the cellular correlate of the mechanism for holding online information." She let her words sink in for a moment and added, "So here you have the neurophysiology of cognition."

There was a long pause, during which both of us stared at the graph. Goldman-Rakic started laughing. "You frown so!" she said. I confessed that I was having a hard time grasping the significance of her work. Of all the topics I had covered as a journalist, I said, neuroscience was the hardest—harder even than particle physics. Goldman-Rakic chortled and called out to a young woman walking through the room, "He's saying that neuroscience is harder than particle physics!" Turning back to me she said, "I'm trying to make it easy for you!"

My problem, I said, was making the transition from these graphs showing the firing rates of neurons to big concepts like memory and cognition and free will. I could understand reductionism when it came to particle physics, but when it came to the human mind, I felt I was missing something. "I want to kill you," she said. "Here I am putting all this energy into explaining this, and you say it's too hard." But surely I wasn't the only person who had ever reacted in this way to her explanations, I replied; philosophers even had a term for this reaction, the *explanatory gap*.

"I think it's in your head that there's an explanatory gap," said Goldman-Rakic firmly. "The moment-to-moment changes in the cells and the brain and all of that are certainly not worked out. And what makes you you, and me me, I'm not going to explain today, and maybe never." Scientists could not understand the origin of the universe either, she said. Nevertheless, she assured me, "we are on the road to understanding human cognition."

Others have sensed an explanatory gap when confronting the research of Goldman-Rakic and her colleagues in neuroscience. Shortly before I left *Scientific American* in 1997, I edited an article on working memory in which Goldman-Rakic was prominently featured. The article's author was another staff writer for *Scientific American*, Timothy Beardsley, a veteran science journalist with a doctorate in animal behavior from the University of Oxford. During the editing process, Beardsley confessed that he had never encountered work more difficult to comprehend and present in a coherent, satisfactory form. He felt as though he was missing something.

Several months after the publication of Beardsley's article, "The Machinery of Thought," *Scientific American* printed a letter that touched on the problem with which Beardsley and I had struggled. The letter complained that the research described in Beardsley's article "tells us only where something happens in the brain, not what the actual mechanisms are for recognizing, remembering and so on. And that, of course, is what we really want to know."

Getting in Touch with Emotions

Even if they unravel the mechanisms underlying working memory and other cognitive functions, neuroscientists must face another problem: How does emotion fit into the puzzle? Until recently many neuroscientists sought to sidestep emotion in their experiments, treating it as an annoying source of experimental noise and distortion rather than a fundamental part of human nature. Neuroscientists have followed the lead of cognitive scientists, who have tried to understand those information-processing functions that can be most easily duplicated in computers, such as vision, recollection, speech recognition, and reasoning.

By avoiding emotion, neuroscientists and cognitive scientists have created a peculiarly one-dimensional picture of the mind, according to Joseph LeDoux, a neuroscientist at New York University. Cognitive science "is really a science of only a part of the mind, the part having to do with thinking, reasoning, and intellect," LeDoux complained in his 1996 book, *The Emotional Brain*. "It leaves emotions out. And minds without emotions are not really minds at all. They are souls on ice—cold, lifeless creatures devoid of any desires, fears, sorrows, pains, or pleasures."

LeDoux, himself a cool, controlled man with deep-set eyes and a carefully trimmed beard, has demonstrated that at least one emotion, fear, can be approached empirically. Unlike language or other cognitive functions unique to humans, LeDoux pointed out, fear is a biological phenomenon whose roots reach back far into the history of life. The neural circuitry and processes that underlie fear have been highly conserved through evolution; thus experiments on rats and other animals may reveal much about humans. The amygdala, which is crucial to the fear response, is found not only in humans and primates but also in rats.

"The fear system is very, very simple," LeDoux told me. "You've got a stimulus that comes in through standard input channels, goes to the amygdala and goes out through the output channels," he said. Early studies of fear responses had produced confusing results because the experiments were too complex. "Every time you change the experiment, you change the way the brain accomplishes the task. So the key in figuring out the fear system is to strip it down to a simpler model."

LeDoux has carried out experiments in which rats have been conditioned to associate a certain sound, such as a musical tone, with an unpleasant sensation, such as an electric shock. The initial response of rats and many other animals to such a stimulus is to freeze, an appropriate tactic for an animal threatened by a predator. The freeze response is an innate, reflexive function. LeDoux and his colleagues showed that damage to a minute structure within the amygdala, called the lateral nucleus, prevented rats from learning to freeze in response to the tone preceding an electric shock. The cognitive ability of the rats was unimpaired in other respects.

LeDoux was trying to unravel the circuitry required for more complex fear-related behavior, which is sometimes called instrumental learning. For example, when a rat learns that freezing does not prevent him from being shocked, he tries avoidance—moving to a different part of the cage or climbing up its sides. At this point, the rat makes the transition from being an emotional reactor to an actor, LeDoux said, capable of making choices and trying different strategies.

Psychologists once believed that the subjective sensation of fear is the first component of the fear response; increased heart rate, sweating, and other physiological symptoms were thought to be triggered by the subjective sensation. LeDoux contended that the opposite is probably true; physiological symptoms occur first and then initiate the subjective sensation of fear. In many cases, moreover, the fear response might never generate a conscious sensation. Our conscious, subjective feelings "are red herrings, detours, in the scientific study of emotions," LeDoux has written.

LeDoux felt that too much attention had been paid to consciousness lately. "It would surely get you the Nobel prize if you figured it out," he told me, "but I don't think it would tell us what we need to know" about the mind. Although consciousness is often equated with the mind, most mental processes occur beneath the level of awareness, LeDoux pointed out. Consciousness, moreover, is a relatively recent innovation of evolution. "Basically the brain is unconscious. Somewhere in evolution consciousness evolved as a module. It's connected up to some other parts of the brain, but not the rest of it."

Explaining consciousness is not as important as understanding how the brain draws on both genes and experience to create a self, a personal identity, in each individual. "That to me is the big question: how our brain makes us who we are. Explaining consciousness wouldn't explain that." The key to this issue is understanding how both nature and nurture affect the brain's wiring. "What's often overlooked is that nature and nurture speak the same language, which is the synaptic language," LeDoux said. Ultimately all influences on personality, genetic or experiential, become manifest at the level of the connections between neurons.

LeDoux doubted whether any single theory would account for emotion. There are many aspects of emotion, he noted. "There's an evolutionary component, there's a cognitive component, a behavioral component. It's just a question of what the balance in the particular situation is." Cognitive theories tend to focus on conscious emotional processes; evolutionary theories emphasize innate emotional responses; behavioral theories stress the role of environmental conditioning. "In any particular emotional episode, it's not a matter of which one is right but which one explains which part of the episode." Moreover, each emotion probably requires a separate explanation; the mechanisms underlying fear are probably quite different from those underlying lust or hatred.

LeDoux summarized the research that he and others have done on emotion, and particularly fear, in *The Emotional Brain*. He also cautiously suggested that investigations of the neurobiology of fear might at some point yield better treatments for human anxiety disorders. LeDoux expected psychiatrists to dismiss his rat experiments as irrelevant to their work. But to his surprise, psychiatrists responded to his book enthusiastically—almost *too* enthusiastically, LeDoux suggested. "It's been almost this uncritical acceptance," he explained. "'Yes, let's go! This is the answer!' They seem so desperate. I don't think I have the answers in my book. I just threw out some ideas."

Like Gerald Fischbach, Torsten Wiesel, and other leading neuroscientists, LeDoux readily acknowledges the shortcomings of his field. He once stated, "We have no idea how our brains make us who we are. There is as yet no neuroscience of personality. We have little understanding of how art and history are experienced by the brain. The meltdown of mental life in psychosis is still a mystery. In short, we have yet to come up with a theory that can pull all this together. We haven't yet had a Darwin, Einstein or Newton."

Then LeDoux suggested that neuroscience might not *need* a unifying theory:

> Maybe what we need most are lots of little theories. It would be great to know how anxiety or depression works, even if we don't have a theory of mental illness. And wouldn't it be wonderful to know how we experience a wonderful piece of music (be it rock or Bach), even in the absence of a theory of perception. And to understand fear or love in the absence of a theory of emotion in general wouldn't be so bad either. The field of neuroscience is in a position to make progress on these problems, even if it doesn't come up with a theory of mind and brain.

Gagian Neuroscience

Neuroscience might find it difficult to produce even LeDoux's "little theories." A fundamental impediment to progress in neuroscience—or in any other mind-related field for that matter—is the enormous variability of all brains and minds. This problem emerged early on from studies of brain-damaged human subjects, who have long provided clues about the links between the brain and the mind. Let's call this research Gagian neuroscience in honor of one of its most famous subjects, Phineas Gage. The twenty-five-year-old Gage was supervising the construction of a railroad line in Vermont in 1848 when an accidental explosion blew an iron bar more than a yard long into his cheek and clear through the top of his head. Not only did Gage live; he remained completely lucid. About an hour later, he was examined by a physician named Edward Williams. Williams recalled that during the examination, Gage "talked so rationally and was so willing to answer questions, that I directed my inquiries to him in preference to the men who were with him at the time of the accident, and who were standing about at this time." A year later another doctor pronounced Gage "completely recovered."

Lofty theoretical edifices were erected upon Gage's injury. For several decades, his case was seen as a setback to the contention of the phrenologist Franz Gall and others that the brain is divided into subsystems dedicated to different tasks, such as speech, movement, and vision. Early examinations of Gage suggested (wrongly, as it turned out) that his brain had been damaged in regions dedicated to language and motor control—and yet these functions remained intact. The conclusion was that the brain is not modular (to use the modern term); rather, it is an undifferentiated mass that functions holistically.

Twenty years after the accident, a physician named John Harlow offered a quite different interpretation of Gage's case. Harlow, who had examined Gage many times over the years, revealed that Gage's personality, if not his functional ability, had changed profoundly after his accident. A previously fastidious, thoughtful, and responsible man, Gage had become "fitful, irreverent, indulging at times in the grossest profanity (which was not previously his custom), manifesting but little deference for his fellows, impatient of restraint or advice when it conflicts with his desires. . . . In this regard his mind was radically changed, so decidedly that his friends and acquaintances said he was 'no longer Gage.'" Gradually Gage's case came to be seen as a corroboration rather than refutation of the modularity hypothesis. The parts of Gage's brain that had sustained the most damage were his frontal lobes, which are now believed to be the seat of such lofty cognitive functions as moral reasoning and decision making.

Gagian neuroscience has supported the view of the mind as an assortment of modules linked to extremely specific functions and traits. Speech disorders caused by brain damage are lumped under the umbrella term *aphasia*. Some aphasics lose the ability to recall the names of people, or of animals, or of inanimate objects. Others can no longer decode verb endings. Aphasics may be able to have a conversation but not to read or write, or vice versa. Brain damage can result in dramatic additions to, rather than subtractions from, a person's psyche. Physicians have reported more than thirty cases of a condition known as gourmand syndrome, in which damage to the right frontal lobe results in an obsession with fine food. A Swiss political journalist made the most of his condition; after recovering from his stroke, he started writing a food column.

A major source of data for Gagian neuroscientists has been patients whose epilepsy is so severe that it can be treated only by severing the corpus callosum, the bundle of nerves connecting the two hemispheres of the brain. (The operation prevents the uncontrolled neural activity that precipitates seizures from engulfing the entire brain.) By studying these patients, the Nobel laureate Roger Sperry and others determined in the 1960s and later that each hemisphere serves different functions. The left hemisphere exerts primary control over language and speech, while the right hemisphere predominates in tasks involving vision and motor skills. The burgeoning field of split-brain research soon spawned the now-familiar pop culture clichés: our left brain embodies our "rational" self and our right brain our spontaneous, "creative" self.

A slew of self-help books—such as *Drawing on the Right Side of the Brain* and *Right Brain Sex*—offered advice on how to escape the confines of our stuffy left brains and become free-spirited right-brain types. Newspapers advertised subliminal tapes that supposedly expanded mental power by delivering different motivational messages to each hemisphere at the same time. Educators proposed revamping curricula to "unleash the right side" of students' brains. Scholars reinterpreted history through the lens of split-brain research; according to one historian, Stalin was a "left-hemispheric leader," and Hitler had a "right-hemispheric temperament."

Even those who should have known better, such as Michael Gazzaniga of Dartmouth University, a pioneer of Gagian neuroscience, contributed to the hype. In his 1985 book *The Social Brain*, Gazzaniga presented a critique of the welfare state based on his interpretation of split-brain experiments. More than a decade later

Gazzaniga was expressing doubts about even some of the most cautious claims concerning the right and left hemispheres. In an article published in *Scientific American* in 1998, Gazzaniga emphasized the hazards of generalizing about the brain based on relatively few cases. People who suffer identical forms of brain damage may exhibit completely different effects. Moreover, the brain's plasticity makes it difficult to reach firm conclusions about the effects of brain damage on even *the same person;* individuals, after all, change over time.

Severe damage to the left hemisphere usually results in permanent impairment of speaking ability, but that was not true of a patient named J.W. Although an operation on his left hemisphere left him mute, J.W. acquired the ability to speak by means of his right hemisphere—thirteen years after his original surgery. A British boy named Alex presents an even more remarkable case. Alex was born with a left hemisphere so malformed that he suffered from constant epileptic seizures. He was also completely mute. When Alex was eight, surgeons removed his left hemisphere to alleviate his epilepsy. Although physicians advised his parents not to expect improvement in his other symptoms, Alex began talking ten months later, and by the age of sixteen he was speaking fluently.

Gagian neuroscience highlights a major obstacle to understanding the human brain. The putative cornerstone of science is the ability to replicate experiments and thus results. But replicability poses a special challenge to mind-science because all brains, and all mental illnesses, differ in significant ways. This lesson emerges quite clearly from the history of lobotomies, according to Jack Pressman, a historian of medicine at the University of California at San Francisco. In his 1998 book, *Last Resort: Psychosurgery and the Limits of Medicine,* Pressman noted how difficult it was to draw firm conclusions about the value of the lobotomy, which treats mental illness by destroying the brain's prefrontal lobes. Some patients seemed to benefit from the procedure; others were devastated. Some patients became wildly uninhibited, like Phineas Gage; others were left virtually catatonic. Pressman concluded: "Because every individual is comprised of a singular combination of physiology, social identity, and personal values, in effect *each patient constitutes a unique experiment.*"

In September 1998, scientists from around the world gathered in Cavendish, Vermont, to commemorate the 150th anniversary of the accident of Phineas Gage. A report on the meeting in *Science* noted that researchers have still not settled the questions originally raised by Gage's case; scientists are "debating whether the frontal cortex functions as a unit or subdivides its duties." One participant in the meeting bravely declared that "the truth probably lies somewhere in the middle."

The Faddishness of Psychology

The skepticism of Socrates about the application of physical theories to human thought and behavior has proved to be extraordinarily prescient. Neuroscience remains peculiarly disconnected with higher-level approaches to the mind, such as psychiatry. The British neurophysiologist Charles Sherrington, who won a Nobel prize in 1932 for his studies of the nervous system, once wrote, "In the training and in the exercise of medicine a remoteness abides between the field of neurology and that of mental health, psychiatry. It is sometimes blamed to prejudice on the part of one side or the other. It is both more grave and less grave than that. It has a reasonable basis. Physiology has not enough to offer about the brain in relation to the mind to lend the psychiatrist much help."

The ascent of psychopharmacology in the 1960s led to hopes that mental disorders could be explained in biochemical terms. Because antipsychosis drugs such as chlorpromazine and reserpine boost levels of the neurotransmitter dopamine in the brain, psychiatrists began to view schizophrenia as a dopamine-related disorder rather than as a consequence of psychic trauma. The advent of antidepressant medications called monoamine oxidase inhibitors and tricyclics, which elevate levels of the neurotransmitters norepinephrine and serotonin, led to speculation that depression stems from a deficit of these neurotransmitters. The growing popularity of selective serotonin reuptake inhibitors such as Prozac has shifted the focus to serotonin alone as the key to depression. (As of yet, there is no accepted explanation for lithium's effect on manic depression.) But even the originators of these neurotransmitter models of mental illness acknowledge their weaknesses. Given the ubiquity of a neurotransmitter such as serotonin and the multiplicity of its functions, it is almost as meaningless to implicate it in depression as it is to implicate blood. Moreover, as Chapter 4 will show, medications for mental illness are not as effective as they are often said to be.

Neuroscientists have sought to find physiological correlates of schizophrenia and other disorders by probing the brains of the mentally ill with PET and other imaging technologies. So far these efforts have yielded frustratingly ambiguous results. Typical was a widely publicized MRI study performed in 1990 at the National Institute of Mental Health. The researchers compared the brains of fifteen schizophrenics to the brains of their nonschizophrenic identical twins. All but one of the schizophrenics had larger ventricles—fluid-filled cavities in the center of the brain—than the nonschizophrenics. Lewis Judd, then the director of the National Institute of Mental Health, hailed the study as a "landmark" that provided "irrefutable evidence that schizophrenia is a brain disorder." Unfortunately, the researchers could not establish whether the enlarged ventricles were a cause or an effect of schizophrenia—or of the drugs used to treat it. Follow-up studies also showed that many normal people have relatively large ventricles, and many schizophrenics do not.

There has also been a troubling schism between neuroscience and psychology. Neuroscientists are "making fundamental discoveries of great importance," the Harvard psychologist Jerome Kagan once remarked. "But the observable behavioral events to which these individual discoveries apply are often unclear . . . the big prize is understanding the relation between molecular and behavioral events. Each domain is moderately autonomous."

This aspect of the explanatory gap was touched on in a 1998

article in *American Scientist*, "Psychological Science at the Cross-roads." The three authors, all psychologists, searched for references to neuroscience in the four most influential psychology journals: *American Psychologist, Annual Review of Psychology, Psychological Bulletin,* and *Psychological Review.* They found that the enormous increase in neuroscience research was not reflected in psychology citations. "Clearly neuroscience is rising in prominence but, according to our measures, not within mainstream psychology."

So far neuroscience has failed to bring about the sort of consensus within psychology that has marked the progress of other fields of biology. The neuroscientists V. S. Ramachandran and J. J. Smythies of the University of California at San Diego recently made this point in an essay in *Nature:*

> Anyone interested in the history of ideas would be puzzled by the following striking differences between advances in biology and advances in psychology. The progress of biology has been characterized by landmark discoveries, each of which resulted in a breakthrough in understanding—the discoveries of cells, Mendel's laws of heredity, chromosomes, mutations, DNA and the genetic code. Psychology, on the other hand, has been characterized by an embarrassingly long sequence of "theories," each really nothing more than a passing fad that rarely outlived the person who proposed it.

One psychological fad, or "theory," that has outlived its inventor is psychoanalysis. Although psychoanalysis has become in certain scientific circles the epitome of pseudoscience, some leading neuroscientists still find Freud's ideas compelling. Susan Greenfield of the University of Oxford is the director of England's Royal Institution and one of England's most prominent neuroscientists. "One of the reasons I admire Freud, above and beyond, perhaps, his specific theories, was that he was a pioneer," Greenfield told a British journalist in 1997. "I am quite unusual, perhaps, as a neuroscientist in finding Freud inspirational."

Actually, Greenfield's affinity for Freud is shared by Floyd Bloom, chairman of the Department of Neuropharmacology at the Scripps Research Institute, author of several books on neuroscience, and editor in chief of the journal *Science*. When I asked Bloom whether he thought neuroscience might one day validate psychoanalysis, he replied, "I don't disagree with that." Twenty years earlier, he told me, he became convinced that neuroscience might provide insights into the abrupt shifts in perspective, or "intellectual gear shifting," that can occur during psychoanalysis. Bloom considered joining a psychoanalytic institute to gather material for his project; he decided against the move only because a sudden advance in molecular biology, which made it possible to mass-produce genes, lured him back to his laboratory.

Another high-profile Freudophile is Gerald Edelman, who won a Nobel prize for his work in immunology, switched later to neuroscience, and now directs the Neurosciences Institute in La Jolla, California. Edelman dedicated *Bright Air, Brilliant Fire,* a popular account of his theory of the mind, to "two intellectual pioneers, Charles Darwin and Sigmund Freud. In much wisdom, much sadness." Edelman remarked in a chapter on the unconscious:

> My late friend, the molecular biologist Jacques Monod, used to argue vehemently with me about Freud, insisting that he was unscientific and quite possibly a charlatan. I took the side that, while perhaps not a scientist in our sense, Freud was a great intellectual pioneer, particularly in his views on the unconscious and its role in behavior. Monod, of stern Huguenot stock, replied, "I am entirely aware of my motives and entirely responsible for my actions. They are all conscious." In exasperation I once said, "Jacques, let's put it this way. Everything Freud said applies to me and none of it to you." He replied, "Exactly, my dear fellow."

Psychoanalysis and Sea Snails

Equally enamored of Freud is Eric Kandel, director of the center for neurobiology and behavior at Columbia University. Kandel has dominated neuroscience for decades through a combination of brilliance and bullying. He is a coauthor of two leading neuroscience textbooks, *Principles of Neural Science* and *Essentials of Neural Science and Behavior*, and he has also exerted an influence on popular accounts of neuroscience. When displeased with the coverage of neuroscience in the *New York Times*, *Scientific American*, or elsewhere, he is known to call editors and reporters to complain and suggest how the coverage can be improved.

Born in Vienna, Kandel was trained in psychiatry at New York University and Harvard, but by the early 1960s he had turned exclusively to neuroscience. He decided to study the nervous system not of *Homo sapiens* but of *Aplysia californica*, a sea snail that has been described as a "purplish-green baked potato with ears." The creature's nerve cells are the largest known to science; they can be seen by the unaided human eye. It was a perfect laboratory for Kandel's investigations into the molecular basis of memory and learning. When sprayed in a certain spot with a jet of water, Aplysia jerks back inside a mantle. When touched repeatedly, however, it withdraws more lackadaisically and finally disregards the stimulus entirely. Through this process, called habituation, the sea snail learns not to associate the jet of water with harm.

Kandel and his colleagues produced the opposite of habituation—an effect called *sensitization*—by repeatedly spraying Aplysia while giving it an electric shock. The animal quickly learns to withdraw at even the slightest touch. Kandel's group showed that both habituation and sensitization produce molecular changes in the neurons controlling Aplysia's withdrawal reflex. In the case of habituation, neurons discharged fewer neurotransmitting molecules into the synapses connecting them to adjoining neurons; sensitized neurons, conversely, discharged more neurotransmitter. These experiments provided evidence for a proposal, first advanced in the 1950s by Donald Hebb, that learning varies the strength of the connections between neurons. This Hebbian mechanism serves as the basis for an artificial intelligence model called neural networks (which I discuss in Chapter 7).

In the 1990s Kandel and his colleagues performed experiments on what has been hailed as a potential "$e = mc^2$ of the mind," a protein that apparently serves as a master switch in the formation of memory. Together with other groups, Kandel's team showed that the CREB protein helps transform short-term memories into long-term ones in Aplysia; when the protein is chemically neutralized, Aplysia cannot form the long-term memories characteristic of sensitization or habituation. (CREB stands for cyclic AMP-response element binding.) Other researchers have performed similar experiments in fruit flies, mice, and other organisms.

At an age when most scientists are content to leave the field to younger colleagues, Kandel has remained very much in the fray. An article on memory research in the *New York Times Magazine* in February 1998 featured a full-page photograph of Kandel wearing a blue-striped shirt and red bowtie and gripping a slime-glazed *Aplysia*. The article noted that Kandel had "pioneered much of the research into the molecular basis of memory" and remained at the forefront of his field. He was trying to parlay his scientific achievements into commercial success by forming a company called Memory Pharmaceuticals, which markets drugs that allegedly decelerate, stop, or even reverse memory loss.

The article mentioned that Kandel had been interested in psychoanalysis before turning to neuroscience. What the article did not mention is that Kandel had undergone psychoanalysis early in his career and even considered becoming an analyst. Although he was "fruitfully distracted by neurobiology" (as the *Times* put it), he never stopped believing in the theoretical and therapeutic potential of psychoanalysis. He retained the hope that Freud's theories about the mind will one day be substantiated by neuroscience.

Kandel spelled out this hope in "A New Intellectual Framework for Psychiatry," published in *American Journal of Psychiatry* in April 1998. He noted that his experiments and others had shown that experience produces physical changes in neurons. More specifically, habituation or sensitization of neurons can turn genes on or off and otherwise affect their expression. These findings implied that experience, such as traumatic events in childhood, could cause neurosis through both neurochemical and genetic effects. In the same way, psychoanalysis and other psychotherapies might produce long-term beneficial effects with a genetic basis. "As a result of advances in neural science in the last several years," Kandel proclaimed, "both psychiatry and neural science are in a new and better position for a rapprochement, a rapprochement that would allow the insights of the psychoanalytic perspective to inform the search for a deeper understanding of the biological basis of behavior."

I met Kandel in the fall of 1997 in his office on the sixth floor of the Psychiatric Institute in Manhattan. His office overlooks the Hudson River, and as we shook hands the sun was descending, blood red, behind the New Jersey skyline. Like other neuroscientists whom I had interviewed, Kandel oscillated between pride and humility as he reviewed his field's performance. When I asked if he thought memory was on the verge of becoming a "solved" problem, Kandel grimaced and shook his head. He noted that the great neuroscientist Ramon y Cajal had once said that problems are never exhausted; only scientists are.

It is possible, Kandel elaborated, that the CREB protein and other findings could reveal the common basis for many different types of memory, just as the discovery of DNA's structure had provided a unified vision of heredity. But memory is "far from solved," Kandel emphasized. Researchers must still determine how the different regions of the brain contribute to the encoding, consolidation, storage, and recall of a memory. "We don't know a goddamn thing about any of those things."

Most scientific fields, he mused, alternate between periods during which they become more complex and periods during which they become more unified. "We are now in an age of splitting," he said. He had had to revise his classic textbook, *Principles of Neural Science,* three times since it was first published in 1981 to accommodate the deluge of new findings. "The easy problems have been solved," Kandel said. "We are now confronting those that are most difficult."

A central problem for neuroscience, he remarked, was learning how the brain constructs pictures of the world from many disparate pieces. The brain does not mirror the world the way a camera does, Kandel emphasized; "it decomposes the image, it decomposes all sensation, and then reconstructs it." Research done by Patricia Goldman-Rakic and others on live primates could yield clues about how the brain creates its picture of reality. "I think that's a very effective methodology," Kandel said. But like Torsten Wiesel and Gerald Fischbach, Kandel emphasized that the binding problem—the Humpty Dumpty dilemma, to use my term—remains very much unsolved.

When he first became a neuroscientist, Kandel recalled, he thought there would be a "rapid merger" between neuroscience and psychiatry. Obviously that synthesis has not occurred. Kandel said that psychoanalysts, who dominated psychiatry in the 1950s and 1960s, were partly to blame for this lack of progress. "Psychoanalysis went through a phase in which it was so confident of its effectiveness that it expanded its interests to all areas of psychiatric disease and all areas of medicine. That was part of its downfall. Insofar as it works, it probably only works in a limited set of circumstances." Psychoanalysts had also been "deliquent" in not questioning their own methods and putting them to the test, Kandel said.

Many of Freud's larger ideas—such as his assertion that childhood conflicts shape character and that much of our mental life occurs below the level of awareness—are now seen as "obvious," Kandel said. "I think everyone sort of accepts that." But questions remain about more specific aspects of Freudian theory, such as the precise manner in which childhood experiences give rise to various personality traits and disorders. "Do these hold empirically and under what circumstances? Are they universal? And even more importantly, does psychoanalysis work and under what circumstances?"

Kandel had a "gut feeling" that psychoanalysis works—his own analysis had made him a better person, he assured me—but *proving* that it works is another matter. Research could show that psychotherapy produces beneficial changes in the brain that "are as specific—maybe more specific!—than drugs. That would be quite wonderful." After all, if talking to a friend or pastor or therapist produces changes in the brain, as it must, "why should that be of any less value than using Prozac, right?"

Even if research cannot demonstrate that psychoanalysis works, it will remain a "very humane, rich perspective on the human mind." Early in this century psychoanalysis served as a much-needed countermeasure to the excesses of behaviorism, which offered a "very shallow" picture of mental representation. Psychoanalysis also anticipated the discovery of modern neuroscience and cognitive psychology that the brain constructs reality rather than simply mirroring it. "So at the worst [psychoanalysis] can give us a *weltanschauung* which is quite rich. At best it may actually turn out to be a therapy which has real utility."

It is possible, Kandel said, that the effectiveness of psychoanalysis may stem from the expectations of patients—in other words, from the placebo effect. "Maybe psychoanalysis is simply a very effective way of recruiting a patient's trust for therapy purposes," he said. "One would hope that there's more, but that could be all of it." Kandel brushed aside the suggestion that such a finding would place psychoanalysis on the same level as faith healing. Faith healers, he asserted, are much more likely to be charlatans and frauds than analysts, psychiatrists, and others closer to the scientific mainstream. "That's not to say that among well-trained physicians you won't find a charlatan, but the statistical probability is much reduced."

Freud as Neuroscientist

Ironically, Freud himself toward the end of his career seemed to doubt whether neuroscience would provide deep insights into the human psyche. Before creating psychoanalysis, Freud spent more than a decade performing research that would now be characterized as neuroscience. He studied the nervous system of lampreys and crayfish, and from 1882 to 1885 he worked closely at the Vienna General Hospital with brain-damaged patients. He published more than three hundred papers and five books on neurobiology, including a monograph on aphasia and other conditions resulting from neural trauma.

In 1895 Freud briefly became convinced that the human psyche and its disorders could be understood in terms of purely physiological phenomena, such as the newly discovered neurons. He wrote to his friend Wilhelm Fliess:

> One evening last week when I was hard at work . . . the barriers were suddenly lifted, the veil drawn aside, and I had a clear vision from the details of the neuroses to the conditions that make consciousness possible. Everything seemed to connect up, the whole worked well together, and one had the impression that the Thing was now really a machine and would soon go by itself.

That same year Freud sketched out his vision of a physiologically grounded theory of the mind in a manuscript that later came to be called *Project for a Scientific Psychology*:

> The intention is to furnish a psychology that shall be a natural science: that is, to represent psychical processes as quantitatively determinate states of specifiable material particles, thus making those processes perspicuous and free from contradiction. Two principal ideas are involved: 1, What distinguishes activity from rest is to be regarded as Q, subject to the general laws of motion. 2, The neurons are to be taken as the material particles.

Freud never published his manuscript, and just a few months later he wrote to a colleague: "I no longer understand the state of mind in which I hatched out the *Psychology*." Immediately after this period he began constructing a purely psychological model of the mind, psychoanalysis. Over the course of his career, Freud became increasingly skeptical about whether the mind and its disorders could be explained in physiological terms. Just before his death in 1939, he seemed to rule out the possibility that psychology would ever be united with neuroscience:

> We know two things about what we call our psyche (or mental life): firstly, its bodily organs and scene of action, the brain (or nervous system) and, on the other hand, our acts of consciousness, which are immediate data and cannot be further explained by any sort of description. Everything that lies in between is unknown to us, and the data do not include any direct relation between these two terminal points of our knowledge. If it existed, it would at the most afford an exact localization of the processes of consciousness and would give us no further help toward understanding them.

Like Socrates more than two thousand years before him, Freud seemed to be suggesting that the explanatory gap might never be closed. His premonition has been borne out so far by the inability of neuroscience either to confirm or to falsify Freud's own theories.

2

WHY FREUD
ISN'T DEAD

. . . .

It is quite possible—overwhelmingly probable, one might guess—that we will always learn more about human life and human personality from novels than from scientific psychology.

—NOAM CHOMSKY, *LANGUAGE AND PROBLEMS*

   OF KNOWLEDGE

On a drizzly spring day in 1996, I sat in a hotel ballroom and listened to scores of Freud's intellectual descendants divulge their fears and desires. The occasion was a meeting of Division 39 of the American Psychological Association. About four hundred members of the division, a haven for devotees of psychoanalysis, had convened at the opulent Waldorf-Astoria in New York City for five days. The meeting's official theme was upbeat: "Psychoanalysis: A Creative Journey." But I had been drawn there by sessions alluding to a darker reality: "The Death of Psychoanalysis: Murder, Suicide or Rumor Greatly Exaggerated?" "Psychoanalytic Technique: Does It Have a Future?" and "Psychoanalysis in Retreat."

The anxiety overflowed at a "town meeting," during which members of Division 39 could discuss any topic they chose. Morris Eagle, president of Division 39 and a prominent New York analyst, opened the meeting by recommending that they address the "primary issue" facing them: "the survival of psychoanalysis in particular, and long-term psychodynamic treatment of any kind, given the realities of managed health care."

Audience members proposed various ways in which they could fight back. Any time someone "knocks" psychoanalysis in public, a woman suggested, a member of Division 39 should respond by writing an article or editorial. The only way to promote psychoanalysis, a man asserted, is to demonstrate empirically its benefits relative to other treatments; in the newly competitive atmosphere engendered by managed health care, "we have to demonstrate we have a better product." Yes, someone added, analysts needed to do research proving that psychoanalysis can reduce medical costs, absenteeism, alcoholism.

Others seemed doubtful. A self-described former mathematician rose to declare that "you can prove anything with statistics, and unfortunately the public is beginning to recognize that." "I don't think we can win on an empirical basis," someone else agreed; the benefits of psychoanalysis can be judged only on a "subjective, existential basis." Eagle, the division president and a Manhattan-based analyst, warned that studies attempting to measure the benefits of psychoanalysis could play into the hands of its opponents.

As the meeting progressed, the tone became increasingly fatalistic. One analyst lamented that his daughter's college catalogue did not list a single course on Freud. Another expressed amazement that psychoanalysis "has managed to get so many people so angry and to get itself so marginalized in such a short period of time." Enemies of psychoanalysis, he grumbled, can be found even within Division 39's own parent organization, the American Psychological Association. Freud himself, someone pointed out, had expressed doubts toward the end of his career that psychoanalysis would survive as a mode of therapy.

A woman from La Jolla, California, revealed that she and other analysts in her region were having increasing difficulty keeping enough patients to stay in business. She complained bitterly that, because of recent court rulings, she could be sued for malpractice if she did not recommend drugs for disturbed patients. "Maybe it's time for me to retire," she sighed. Many of her colleagues nodded and murmured in gloomy assent.

Some paranoids, the old joke goes, really do have enemies. From the moment Freud began propounding his theories a century ago, his work has been subjected to unrelenting attacks. In 1896 Freud's brand-new theories about the sexual roots of hysteria were derided as "a scientific fairy tale." Four years later a member of the Vienna Medical Society mocked Freud in a skit: "If the patient loved his mother, it is the reason for this neurosis of his; and if he hated her, it is the reason for the same neurosis. Whatever the disease, the cause is always the same. And whatever the cause, the disease is always the same. So is the cure: twenty one-hour sessions at 50 Kronen each."

A 1913 review of what many consider to be Freud's greatest work, *The Interpretation of Dreams*, found in it "a total lack of the characteristics which lead to scientific advance." In 1916 *The Nation* complained that psychoanalysis was "well founded neither theoretically nor empirically," and that same year the periodical *Current Opinion* likened Freud's "sex theory" to "the green cheese hypothesis of the composition of the moon." The Russian novelist Vladimir Nabokov called Freud a "witchdoctor" and "Viennese quack." Nabokov decried "the vulgar, shabby, fundamentally medieval world of Freud with . . . its bitter little embryos spying, from their natural nooks, upon the love life of their parents."

Attacks on Freud reached a crescendo during the 1990s, as authors of books such as *Freudian Fraud, Why Freud Was Wrong, Freud Evaluated*, and *Unauthorized Freud* attempted to drive a stake through Freud's heart. In 1995 the Library of Congress postponed a long-planned exhibit on Freud after a coalition of protesters—including Freud's own granddaughter, Sophie—complained that it was too adulatory. When the exhibit finally opened in the fall of 1998, its catalogue included contributions from several leading Freudo-phobes. One was the British historian Frank Cioffi, who compared belief in psychoanalysis to belief in the Loch Ness monster.

Market forces have inflicted heavy damage on psychoanalysis. Few people have the time or money for a treatment that calls for as many as five one-hour sessions a week at $100 each and typically lasts years. Many patients, and all health insurers, favor short-term therapies that target specific problems rather than delving deeply into a patient's past. Meanwhile, psychiatrists and other M.D.s are increasingly prescribing drugs rather than talk therapy alone for such common ailments as depression and anxiety. Given all these trends, it seems fair to ask, as *Time* magazine did in a cover story in 1993, "Is Freud Dead?"

Hardly. If Freud were truly dead, why would so many critics still be expending so much energy trying to kill him? The answer, of course, is that Freud still has legions of defenders; for every book attacking Freud there is another taking his part. "Freud's recent critics will do him no lasting damage," Paul Robinson, a historian at Stanford University, predicted in *Freud and His Critics*. "At most they have delayed the inevitable process by which he will settle into his rightful place in intellectual history as a thinker of the first magnitude." Freud's influence is particularly strong in the humanities and social sciences. One survey of literature in these realms found that only Lenin, Shakespeare, Plato, and the Bible are cited more often than Freud.

But Freudophilia also infects the kind of scientists who are supposed to know better. To be sure, references to psychoanalysis have declined in mainstream psychology journals over the past few decades, according to a 1998 survey. But the three psychologists who conducted the survey asserted: "This does not mean that 'Freud is dead,' but rather that his presence is felt indirectly. Indeed, many of Freud's basic ideas—for example, that unconscious processes influence behavior and that early-childhood experiences influence adult development—have become incorporated into the foundation of psychology as a science." Even scientists who profess skepticism or indifference toward Freud still use him as a benchmark for judging and explaining newer ideas. Books such as *The Emotional Brain* by Joseph LeDoux, *How the Mind Works* by the cognitive scientist Steven Pinker of the Massachusetts Institute of Technology, and *Searching for Memory* by the Harvard psychologist Daniel Schacter are crammed with references to Freud.

Membership in the American Psychoanalytic Association, the largest analytic society in the United States, has remained surprisingly steady over the past decade at about three thousand, and enrollments in training institutes are rising. The International Psychoanalytic Association includes more than nine thousand members, and it reports that membership is growing in South America, Europe, and elsewhere. In 1996 the Russian president Boris Yeltsin signed a decree recognizing psychoanalysis, which Stalin had banned along with all of Freud's writings in 1930, as a legitimate psychiatric treatment.

So the real question is this: Why *isn't* Freud dead? Richard Webster, a leading Freud basher, provided an answer in his 1995 book, *Why Freud Was Wrong:* "No negative critique of psychoanalysis, however powerful, can ever constitute an adequate refutation of the theories which Freud put forward. For in scientific reality bad theories can only be driven out by good theories." (Webster went on to predict that Darwinian psychology, which I discuss in Chapter 6, will deliver us from Freudian psychology.) Psychoanalysis persists because science has been unable to deliver an obviously superior theory of and therapy for the mind. That is why Freud isn't dead.

Goats, Sheep, and the Oedipal Complex

That is not to say that Freud and his descendants do not deserve to be criticized. One of the earliest and still most legitimate objections to psychoanalysis is that it is almost infinitely flexible; it can account for virtually any observation. Freud's ability to explain away evidence that contradicted his theories was at times almost comical. He often insisted, for example, that neuroses are "without exception disturbances of the sexual function" traceable back to childhood. A serious challenge to this assertion arose during World War I, which left thousands of soldiers suffering from what

gued that these psychological disturbances, clearly triggered by
the trauma of combat, contradicted Freud's claim that all neurosis
was sexual in origin. Freud rebutted this "frivolous and premature" argument in *An Autobiographical Study*, published in 1924. He
asserted that soldiers suffering from battle fatigue were narcissists; they were driven mad during war by the threat to their primary love object, themselves.

Freud's methods for gathering "clinical evidence" were also
questionable, to put it mildly. A key tenet of psychoanalysis is that
the mind represses memories of traumatic childhood episodes,
whether real or imagined. The analyst's job is to intuit what those
episodes were, based on his interpretation of the patient's dreams,
word associations, and other "data" welling up from the unconscious. Only when the patient confronts these memories will the
healing begin. According to Freud, the more traumatic and
important an episode is, the more likely a patient is to resist the
analyst's interpretation. "The amount of effort required of the
physician varied in different cases; it increased in direct propor-
tion to the difficulty of what had to be remembered."

Let's say that an analyst claims, based on his interpretation of a
teenage patient's dream, that the boy exhibits a classic Oedipal
syndrome: he wants to kill his father and copulate with his mother.
If the boy agrees with the interpretation, fine. If he rejects the interpretation, his denial provides even *stronger* proof that he is repressing Oedipal urges. Freud and his followers adopted this same
strategy—which has been described as "heads I win, tails you
lose"—as a defense against all doubters of psychoanalysis. Critics
of psychoanalysis were obviously displaying symptoms of repression and denial.

To understand why Freud has infuriated so many feminists, one
need only read "Some Psychical Consequences of the Anatomical
Distinction Between the Sexes." (Ironically, the lecture was delivered not by Freud, who was ailing, but by his daughter Anna.) The
first time a girl sees a male penis, Freud contended, she "knows
that she is without it and wants to have it. . . . Even after penis-envy

has abandoned its true object, it continues to exist: by an easy displacement it persists in the character trait of *jealousy*." Women who derive sexual pleasure from their clitorises, whether during masturbation or intercourse, have failed to move beyond penis envy and thus to embrace their true feminine nature. The "elimination of clitoridal sexuality," Freud announced, "is a necessary precondition for the development of femininity."

Descendants of Freud, albeit not Freud himself, also promulgated theories of mental illness that demonized mothers. As recently as the 1970s, many psychoanalytically oriented psychiatrists still blamed autism and schizophrenia on "refrigerator" mothers, who withhold affection from their children. Psychoanalysts also warned that mothers could harm their children by being *too* affectionate. "Women have never fared well at the hands of Freudians," the journalist Edward Dolnick commented in *Madness on the Couch*, a recent critique of Freudian psychiatry.

None of the attempts to defend Freud on scientific grounds is very persuasive. One recent example is the 1997 book *The Talking Cure*. Its author, Susan Vaughan, is a psychiatrist at the New York State Psychiatric Center and a practicing psychoanalyst; she sees patients in a Manhattan neighborhood so dense with analysts it is known as the "mental block." Vaughan has received a grant from the National Institute of Mental Health to study how psychotherapy affects the brain. In *The Talking Cure* Vaughan granted that not all of Freud's ideas had fared well. His ideas about female sexuality, in particular, were "culture-bound and outmoded."

Freud was nonetheless "a genius," Vaughan declared, whose discoveries "make better sense today because of what we have learned about the brain." Neuroscience had provided "solid scientific evidence" that both childhood experience and psychoanalysis alter "the way in which the neurons in the brain are connected to one another. This rewiring leads to changes in how you process, integrate, experience, and understand information and emotion." In defense of this thesis, Vaughan cited various experiments. In one, MRI showed that cognitive-behavioral therapy and Prozac produced similar changes in the brains of obsessive-compulsive patients. Vaughan was even more impressed with Eric Kandel's demonstration that learning produces chemical changes in the neurons of sea snails.

Vaughan's reasoning hinges on a rather obvious fallacy. Neither the MRI study of obsessive-compulsive humans nor Kandel's work on snails has any bearing on psychoanalysis per se. These and other experiments do indeed imply that both childhood experiences and psychoanalysis can alter "the way in which the neurons in the brain are connected to one another." So what? One could substitute "taking a course in Chinese cooking" or "watching the Super Bowl" for "childhood experience" or "psychoanalysis" and the statement would be just as true. *Of course* childhood experiences and psychoanalysis during adulthood cause changes in the brain. So does experience in general. That is a truism that no one disputes. But the research Vaughan cited certainly does not establish the validity of psychoanalysis as a theory of or therapy for the mind.

Even Eric Kandel found Vaughan's book unconvincing. Vaughan simply assumed that psychoanalysis works, Kandel complained during my interview with him, and then used his work to justify her belief. His work on sea snails, while suggestive, is hardly a ringing endorsement of psychoanalysis, Kandel said. *Someday* neuroscience may show how psychoanalysis produces beneficial alterations in the brain, he elaborated, but that connection is far from established *now.* "You can't *assume* that this happens. You have to *show* this is happening."

What scientists have established is that Freud had an uncanny ability to concoct theories that can be neither verified nor falsified once and for all through empirical means. Many admirers of Freud consider *The Interpretation of Dreams* to be his greatest literary and scientific achievement. Freud himself said as much in the preface to the third edition of the book: "Insight such as this falls to one's lot but once in a lifetime." Freud hypothesized that during sleep, disturbing desires and fears rooted in childhood experience emerge from the unconscious, albeit still disguised by the ever-vigilant ego. By decoding these images, the analyst could glean more directly the nature of the unconscious.

Various researchers claim to have debunked Freud's theory of dreams. In 1998, researchers at the National Institutes of Health and the Walter Reed Army Institute of Research reported in *Science* that they had scanned the brains of sleeping subjects with PET. The scans showed that the brain's prefrontal lobes, the seat of the brain's highest cognitive functions, are quiescent during the so-called REM phase of sleep, which is marked by rapid eye movement and vivid dreaming. Noting that the prefrontal lobes are the most plausible site of the Freudian ego, the investigators suggested that their work undermined Freud's notion that dreams represent raw, libidinal impulses screened and transformed into cryptic symbols by the ego. Dreams may instead be merely the result of signals that the brain employs to help it determine when it has had enough sleep.

The *New York Times* ran a story on these experiments under the headline, "Was Freud Wrong? Are Dreams the Brain's Start-Up Test?" Six days later the *Times* published letters from de-debunkers. Howard Shevrin, a psychologist at the University of Michigan, pointed out that the report in *Science* had confirmed that dreams draw heavily on dreamers' emotions and long-term memories. This finding "moves neuroscience closer to Freud's theories," Shevrin contended. The researchers' claim that "Freud was wrong," a psychoanalyst chimed in, "reflects bias, not science."

Empirical studies of the Oedipal complex have been similarly inconclusive. Cross-cultural studies have suggested that far from wanting to copulate with their mothers, boys usually have no sexual desire for female family members—and in fact for any other female with whom they are raised in close proximity. According to evolutionary theorists, this aversion does not reflect repression, as Freud proposed; it was instilled in our ancestors by natural selection. Incest, after all, often produces genetically impaired offspring.

What is one to make, then, of "Mothers Determine Sexual Preferences," an article published in *Nature* in 1998? In the article, a group of British and South African scientists described experiments in which newborn goats were raised by female sheep and newborn lambs were raised by female goats. The investigators found that the young males—but not females—tended to mimic the play and grooming behavior of their "foster mothers." When the males matured, they also preferred to socialize—and copulate—with females who resembled their foster mothers rather than biological mothers. In other words, the male goats wanted to mate with female sheep, and the male sheep wanted to mate with female goats. The study "indirectly supports Freud's concept of the Oedipus complex," the authors concluded.

One of the most ambitious attempts to evaluate Freud's scientific merits is the 1996 book *Freud Scientifically Reappraised*. The psychologists Roger Greenberg and Seymour Fisher of the State University of New York in Syracuse assessed Freud's oeuvre based on their analysis of more than eighteen hundred studies published over a period of more than sixty years. They pointed out that Freudian psychology is often viewed as a monolithic entity that must be accepted or rejected in its entirety. In fact, Freud proposed many hypotheses, which were not necessarily dependent on each other; some have withstood scrutiny, and others have not.

That was the explicit theme of *Freud Scientifically Reappraised*. The implicit theme was that it is difficult either to disprove or to confirm any Freudian propositions. That became clear when I spoke to Greenberg in 1998. (Fisher died in late 1996.) For example, Greenberg told me that he and Fisher had found evidence both for and against the notion of repression. Research on implicit memory and related phenomena has established that "there are things that go on out of peoples' awareness, and that do affect the way they react and behave," Greenberg explained. Moreover, "people do try to shut out some of their unacceptable feelings"—for instance, homosexual impulses. But rarely do people completely repress memories of traumatic experiences, as Freud had claimed, Greenberg said. Moreover, research by various psychologists—notably Elizabeth Loftus of the University of Washington—had demonstrated that it is all too easy for therapists to implant false memories in patients. "People are open to a great deal of influence in the psychotherapy situation," Greenberg elaborated.

One aspect of Freud's work that *has* fared well, according to Greenberg, is the categorization of personalities into anal and oral types. "There has been some fairly decent research suggesting that those personality types and the traits that he associated with them seem to hold up when you look at the research evidence," Greenberg said. Anal traits such as obstinacy, parsimony, and orderliness "seem to occur together in the same people, and they do seem to be related to anal concerns." Freud had alleged that parents foster these traits in their children by subjecting them to excessively early or strict toilet training.

But just how reliable are the studies linking toilet training to anal characteristics in adults? In his 1992 book, *Freudian Fraud*, the psychiatrist E. Fuller Torrey examined studies of Freud's anal hypothesis, including some cited by Greenberg and Fisher. Most of the studies presented no data on the toilet training of the subjects, and those that did often found no correlation between the severity of toilet training and anal characteristics. Many of the subjects, notably psychology students, had been exposed to Freudian concepts and might have known what answers were expected of them.

The weaknesses were exemplified by a study performed by Seymour Fisher himself in 1970. Fisher gave students several questionnaires, including a "Body Focus Questionnaire" designed to determine which parts of their own bodies the students were most conscious of. The students also took the so-called Blacky personality test, which consists of pictures of a dog named Blacky shown in situations fraught with psychoanalytic significance. In one scene Blacky sees another dog having his tail chopped off, and in another Blacky defecates between the doghouses of his parents.

Fisher reported that students with a high degree of "back awareness" (presumably including the backside) exhibited more "sensitivity to stimuli with anal connotations, negative attitudes toward dirt, [and] measures of self-control." As E. Fuller Torrey commented drily in *Freudian Fraud:* "No information on toilet training was obtained. It would appear, therefore, that Fisher merely verified a cluster of personality traits consistent with the 'anal character' and that college students with these traits had a higher 'back awareness,' possibly because they were aware of Freudian theory."

Crews Missiles

If Freud is not dead yet, it is certainly not for lack of effort on the part of Frederick Crews. A professor of English at the University of California at Berkeley, Crews first voiced skepticism toward psychoanalysis in the early 1970s. His views did not attract widespread attention until 1993 and 1994, when he attacked Freud in two blistering articles: "The Unknown Freud" and "The Revenge of the Repressed." Crews's polemic was all the more shocking because it appeared in the *New York Review of Books*, which was viewed as a bastion of psychoanalytic thought. (Cartoon drawings of Freud still adorn the journal's subscription cards.)

Crews was not content to repeat well-worn arguments that psychoanalysis is devoid of scientific or therapeutic merit. He contended that Freud was guilty of "dishonesty and cowardice"—and worse. Crews revealed that in the 1920s, Freud diagnosed one of his American followers, a psychoanalyst named Horace Frink, as a latent homosexual. Freud recommended that Frink counteract this tendency by divorcing his wife and marrying a wealthy heiress named Angelika Bijur, with whom Frink was having an affair. Freud simultaneously ordered Bijur to divorce her husband and marry Frink. Freud's real motive, which he revealed in his letters, was to extract contributions from Bijur. Frink and Bijur followed Freud's instructions—divorcing their spouses and marrying each other—with devastating consequences. The divorced spouses died shortly after being abandoned. Bijur then divorced Frink, who sank into a psychotic depression. "It is not recorded whether Freud ever expressed regret for having destroyed these four lives,"

Crews remarked, "but we know that it would have been out of character for him to do so."

Crews's harshest charge was that Freud inspired what is sometimes called recovered-memory therapy (critics call it false-memory therapy). It has been estimated that as many as one million patients in the United States have undergone some type of recovered-memory therapy. Recovered-memory therapists typically contend that vast numbers of children are physically and sexually abused by parents and other adults; although these children repress memories of the abuse, they often display psychological distress as adults. The therapists help the patients achieve a cathartic relief from their troubles by excavating these repressed incidents of abuse.

As recovered-memory therapy proliferated in the 1980s, thousands of patients, most of them women, began accusing parents and other adults of abuse. Some of these cases led to the trial and conviction of the alleged abusers, even when there was no evidence to corroborate the recovered memories. The recovered-memory movement foundered as patients made increasingly lurid and incredible charges—involving satanic rituals, orgies, human sacrifices, and even extraterrestrials. Psychologists specializing in memory testified that there was little or no evidence that memories could be repressed for decades and then exhumed in pristine form. Many patients recanted their accusations and complained that their memories of abuse had been implanted in them by their therapists. Some successfully sued their therapists for malpractice.

Crews acknowledged that blaming this disturbing modern phenomenon on Freud might seem unfair at first glance. In 1896, Freud briefly proposed his so-called seduction theory, which posited that women suffering from hysteria had been sexually molested during childhood by their fathers or other adults; their hysteria was allegedly the result of their repression of these traumatic memories. Freud then retracted this hypothesis, proposing instead that his patients had only *imagined* being molested as children; their mental illness derived from their repression of these illicit, Oedipal fantasies. Thus was psychoanalysis born.

If Freud explicitly rejected sexual abuse as a primary cause of psychic troubles, how could Crews blame him for the modern recovered-memory movement? The answer, according to Crews, is that recovered-memory therapy depends on several Freudian concepts. Both Freud's original seduction theory and his Oedipal theory rested on the same assumption: that many patients are repressing memories of either fantasized or genuine sexual incidents from childhood. But according to Crews, *none* of Freud's patients independently presented Freud with such memories. They did so only after prompting and in some cases browbeating by Freud.

Crews backed up this assertion with damning quotations from Freud. Discussing his patients' recollections of sexual molestation, Freud stated in 1896: "Before they come for analysis, the patients know nothing about these scenes. . . . Only the strongest compulsion of the treatment can induce them to embark on a reproduction of them." Freud wrote elsewhere: "The principal point is that I should guess the secret and tell it to the patient straight out." Crews concluded: "By virtue of his prodding, *both before and after* he devised psychoanalytic theory, to get his patients to 'recall' nonexistent sexual events, Freud is the true historical sponsor of 'false memory syndrome.'"

Crews's double-barreled blast at Freud in the *New York Review of Books* provoked more mail than any previous articles in the journal's history. One observer commented:

> The *New York Review of Books* had come to be regarded by many as the house magazine of a particular section of the American liberal intelligentsia who were deeply sympathetic toward psychoanalysis. That Frederick Crews's critique of Freud should be featured so prominently there was what really hurt. Had Tom Paine been invited to preach at Canterbury Cathedral, and Voltaire been summoned by the Pope to celebrate mass at the

Vatican, the sense of violated sanctity among the faithful could scarcely have been greater.

I first encountered Crews in the lobby of a hotel near the campus of Yale University, where he would participate in a public symposium on Freud the next day. He was tall and slim; his wire-rim glasses gave him an ascetic air. He was dressed like an executioner: black raincoat over a black collarless shirt, black pants, black shoes. As we chatted over dinner in the hotel restaurant, Crews spoke for the most part in a calm, monotonic voice. But his superficially reserved, even shy manner concealed a ferocious resolve and self-assurance. No matter what objection I raised, Crews parried with a tightly reasoned response.

When I suggested that Crews did not fully acknowledge the degree to which psychoanalysis continues to exert an influence on science and the rest of culture, he replied, "On the contrary, I do acknowledge it, and that's why I still think it's worth critiquing. If it weren't widely influential, then I could just shut up." Nevertheless, Crews maintained, an examination of citations in scientific journals and elsewhere would reveal that psychoanalysis is undergoing a "steeply downward trend."

Crews scoffed at the notion that neuroscience and cognitive science had validated Freud by proving the existence of unconscious, or implicit, memory. Someone who is driving a car does so more or less without conscious deliberation, Crews acknowledged. "In the same sense, much of our mental life takes place below the surface of consciousness. But that is in no way a validation of the Freudian dynamic unconscious, which is supposedly seething with repressed desires and fears and traumatic memories."

Evidence for the Freudian unconscious was tenuous at best, Crews said. In the case of the Freudian slip, for example, "Freud's immediate assumption is that it springs from the depths of the unconscious, and it's either aggressive or sexual in nature." A much more plausible explanation of such slips had been proposed by an

Italian scholar named Sebastiano Timpanaro, who was an authority on the corruption of ancient texts. Timpanaro had showed through meticulous scholarship that scribes copying ancient texts often inadvertently introduced errors into their copies by substituting common words for exotic ones. Timpanaro called this process "banalization."

"If something is a little unusual in the original," Crews explained, "the error banalizes it, makes it more normal, familiar to the world of the translator. One banalization can lead to another until you get a highly corrupt text." Timpanaro contended that Freud's own examples of misstatements can be easily explained by banalization. By ignoring this commonsense alternative, Crews said, Freud violated Occam's razor, the principle that the best explanations are those with the fewest assumptions. Freud "never, *ever* does that," Crews said. "In his complete writings, there is no case where he goes for the obvious explanation."

Crews was unimpressed by the argument that no scientific theory can meet the standards he applies to psychoanalysis. "My objections to psychoanalysis are garden-variety objections," he said. Psychoanalysis "justifies its ideas by reference to its own interpretations. If you want to justify the idea of the ego, you justify it by interpreting what a patient says in the light of ego theory. There isn't any reputable science on the face of the earth that's conducted that way."

Crews was at his most cutting on the subject of recovered-memory therapy, which he called "malpractice on a criminal scale," "absolute witch-doctor stuff," and "total quackery." If he and others did not criticize the recovered-memory movement and reveal its Freudian origins, Crews feared, it might keep recurring. "Unless we can challenge the base assumptions, we're going to get the same thing all over again."

I asked why Crews, a self-described atheist, focused his ire on psychoanalysis rather than religion, which had arguably caused much more harm throughout history. Crews granted that religious belief has often led to intolerance, to "fanatical crusades and pogroms." Nevertheless, he added, "I wouldn't want to say that religion at all times and in all places has been a negative influence. Obviously anything that makes a human society cohere without great harm to other people is a benefit." Religion can promote social cohesion and morality in communities that sorely need these qualities, such as inner cities, he said. "Every society lives by myths," Crews said, "and if it completely runs out of myths, it's in big trouble." He murmured, as if to himself, "I personally have run out of myths." Pause. "I hope I have."

The next day, I assembled along with several hundred others in a large, wood-paneled lecture hall at Yale to watch Crews do battle with pro-Freudians. Crews began his lecture by playing off the title of the conference: "Whose Freud? The Place of Psychoanalysis in Contemporary Culture." "To the question posed in our conference title," Crews declared, "I can offer a simple reply: he's all yours. Take my Freud—please! But do you really want him—the fanatical, self-inflated, ruthless, myopic, yet intricately devious Freud who has been unearthed by independent scholarship of the past generation—or would you prefer the Freud of self-created legend, whose name can still conjure the illusion that 'psychoanalytic truth' is authenticated by the sheer genius of its discoverer?"

As Crews continued in this manner for another twenty minutes, the audience responded with several hisses and one low whistle, but a few laughs too. Meanwhile, another panel member, Robert Michels, a psychoanalyst and professor of psychiatry at the Cornell University School of Medicine, provided a mimed commentary, smirking, raising his eyebrows, rolling his eyes, and shaking his head in response to Crews's comments. When it was his turn to speak, Michels informed the audience that as a psychiatrist he did not *care* whether psychoanalysis had been scientifically validated; his only concern was for the welfare of his patients. Michels knew through personal experience that psychoanalysis helped patients, and that was enough for him.

The two other speakers in Crews's session, Judith Butler, a professor of literature at the University of California at Berkeley, and Juliet Mitchell, a psychoanalyst and lecturer at the University of Cambridge, each expressed discomfort with Crews's approach.

Butler suggested that theories about human experience required different methods for verification than theories addressing nonpsychological phenomena. Crews's insistence that scientific theories be supported by empirical evidence, Mitchell complained, would lead to an "impoverishment" of science. Both Butler and Mitchell also seemed intent on making the point that modern psychoanalysis should help society to become tolerant of a broader range of possible relationships, including incestuous ones between parents and children. At least I think this was their point; both of them spoke in almost parodically obscure jargon.

"The prohibitions that work to prohibit non-normative social exchange," Butler said at one point, "also work to institute and control the norms of presumptively heterosexual kinship, where positions such as mother and father are differential effects of the incest taboo. Some analysts treat these positions as if they are timeless and necessary positions, psychic placeholders that every child has or acquires through the entry into language. This is, I think, to miss the point that kinship is a contingent social practice, and that there is no symbolic position of mother and father that is not precisely the idealization and ossification of contingent cultural norms."

During the discussion period, Butler turned toward Crews and asked him with a sly smile why in his opening remarks he had said, "Take *my* Freud." Why not *your* Freud? Her implication was obvious: Crews had committed a Freudian slip. Through his word choice, he had inadvertently revealed the repressed, Oedipal attachment to Freud that motivates his conscious hostility. Crews responded to Butler with a sly smile of his own. Actually, he replied, he had merely been paying tribute to Henny Youngman, who had just died a few weeks earlier and had been one of Crews's favorite comedians. Youngman's trademark joke was, "Take my wife. Please."

Steven Hyman's Skepticism

Crews's view of Freud and psychoanalysis is for the most part compelling. I have yet to see a persuasive rebuttal of what I consider to be Crews's most damaging charge: that Freud invented the "clinical evidence" that led him to propose the Oedipal complex and other key components of psychoanalysis. Certainly none of the pro-Freudian panel members at the Yale conference countered Crews successfully; nor did the respondents to his articles in the *New York Review of Books*.

The meeting at the Waldorf-Astoria that I attended in the spring of 1996, which I described at the beginning of this chapter, also corroborated Crews's linkage of psychoanalysis and recovered-memory therapy. In one packed session, five female therapists discussed "Individual and Marital Treatment of an Incest Survivor." One speaker addressed recent charges that some therapists were implanting "false memories" of horrific childhood abuse in patients. As the audience cheered and applauded, the speaker called these charges "reactionary" and "misogynistic"; those making these claims were obviously carrying out a "political agenda behind the thin veil of science."

She admitted that she was concerned by the rather fantastic nature of some patients' memories, particularly those involving ritual human sacrifice and other horrors for which no physical evidence had ever been found. How should these claims be treated? the speaker asked. It is not the therapist's job, another panel member responded, to determine whether the abuse recalled by a patient actually occurred. "We can only know what is emotionally true at the moment," she said. Trying to develop a "coherent picture" of a patient's history "can be dangerous and limiting." Creating an "emotionally safe environment" for the patient "takes precedence over everything else."

This same disregard for truth characterized other sessions too. One speaker gave a talk whose central theme, as far as I could tell, was that analysts must recognize that they can never really understand their patients, because the human mind "is not a unitary phenomenon" but is "nonlinear." The analyst "is working in a complex field of shifting realities," and thus his attitude toward a patient at any given point "is always inherently both right and wrong." By accepting that he cannot arrive at a true understanding of a patient, the analyst "puts the patient's growth potential back in his own hands."

Ironically, Freud himself forcefully rejected this postmodern attitude toward truth (whatever that is). In *The Future of an Illusion*, Freud denied the "radical" claim that "science can yield nothing but subjective results, whilst the real nature of things outside ourselves remains inaccessible." Of course, Freud's method for gathering evidence for his theories—*heads I win, tails you lose*—undermined his own claims to objectivity.

My main complaint with Frederick Crews is that he exhibits a kind of tunnel vision. He is so fixated on the flaws of psychoanalysis that he refuses to see anything of merit in it. Nor does he situate psychoanalysis within its larger scientific context. The limitations of his critique are most apparent when he attempts to explain why psychoanalysis, given its obvious shortcomings, has remained influential even among hard-nosed scientists. More specifically, why do Eric Kandel, Gerald Edelman, and other prominent neuroscientists still hold psychoanalysis in high regard? When I put this question to Crews, he speculated that they had been brainwashed by undergoing psychoanalysis themselves. He likened psychoanalysis to a religious cult that excels at nothing so much as self-perpetuation. "Those who complete Freudian therapy end up as Freudians," Crews said. "They don't end up as cured patients. They go out like body snatchers and get more Freudians."

This response is inadequate in two respects. First, it does not acknowledge the extraordinary allure of Freud's writings (which I will address shortly). Second, it does not address the imperfections of all the alternatives to psychoanalysis. The psychologists Roger Greenberg and Seymour Fisher made this latter point in *Freud Scientifically Reappraised*. They conceded that psychoanalysis has not met the standards of evidence demanded by some critics.

But anyone who adopts such a "superperfectionistic perspective," they contended, would have to concede that "no psychological theories have been adequately tested." Proponents of a biological rather than psychological view of mental illness, who were among the fiercest critics of Freud, had certainly not demonstrated the superiority of their approach, according to Greenberg and Fisher. "It would be premature to translate the current wishful dreams of glory of biological psychiatry into an agenda that banishes psychodynamic schemata."

If psychoanalysis is an imperfect and unproven mode of mindscience, in other words, so are all its would-be successors. It is not just psychologists such as Greenberg and Fisher—or journalists such as myself—who adhere to this position. Much the same view has been espoused by Steven Hyman, one of the most powerful figures in modern biological psychiatry. In 1996 Hyman, a psychiatrist and neuroscientist at Harvard, was appointed director of the National Institute of Mental Health, where he oversees a research budget of more than $800 million. Shortly after his appointment, I met him at the annual meeting of the American Psychiatric Association in New York.

With his trim beard and strong jawline, Hyman faintly resembles a young Freud, but he is no Freudian. While impressed with some psychoanalytic literature, he found psychoanalytic explanations of the mind ultimately unsatisfying. "What they were writing was a *tour de force*. It made sense. It gave explanations to phenomena that demanded explanations. But I always worried that they were nothing but good stories." On the other hand, some "simple pharmacological models" of mental illness also struck Hyman as mere "stories." "I'm an equal opportunity skeptic," he said. Proponents of such models, he explained, often describe depression as a chemical disorder that can be treated with drugs such as Prozac, which boosts levels of the neurotransmitter serotonin in the brain. "And I'll say, 'What does that *mean*? Is that an answer?'"

The reductionist methods of molecular biology and neuroscience had proved to be extremely effective in recent decades,

Hyman said. "We are living in an age when there is extraordinary excitement being generated by these advances in molecular and cellular neurobiology." Nevertheless, he added, "we're not going to clone our next serotonin receptor and say, 'Now I understand how the brain works.'"

Hyman had a similar attitude toward the field of behavioral genetics. Through studies of identical twins and other research, behavioral geneticists attempt to quantify the relative contribution of genes and the environment to various human traits and disorders. "It's even too simplistic for a person to say what percentage of a given behavioral trait is genes and what is environment," Hyman commented, "because the given environment could actually change the relative contribution." Finding the genes that create susceptibility to schizophrenia and other diseases would represent a tremendous advance, Hyman said. But he warned that researchers would still have to determine how such genes "interact with each other and the environment in building a brain that leads to schizophrenia. And I take it that that is not something that's going to be very easy to solve."

As for the field of evolutionary psychology, which attempts to explain the mind with Darwin's theory of natural selection, Hyman found it fascinating but frustrating, much like psychoanalysis. Evolutionary psychologists often "pooh-pooh in a way that is reminiscent of psychoanalysis their inability to do experiments." There is "a grave danger," Hyman added, in viewing the mind primarily as an instrument designed by natural selection for helping us propagate our genes. As an empirically oriented, "wet biologist," Hyman said, he was "impressed by how jury-rigged" organisms are. Evolutionary psychologists "don't have adequate respect for the way in which the prior history of building our brains may have constrained adaptation, or may have constrained the material which could be selected."

It is not surprising that all these approaches fall short, Hyman said, given that "understanding how the brain works and how things go wrong is the most difficult enterprise that humanity has ever undertaken." The persistent debate over Freud's ideas, Hyman suggested, is an indication of the immaturity of psychology, psychiatry, neuroscience, and other mind-related fields; after all, the field of infectious disease is not still riven by debates between Pasteurians and anti-Pasteurians. In mature scientific fields, Hyman said, investigators rarely examine findings more than five years old.

Was Hyman confident that eventually the mind and its disorders would be fully understood? He grimaced as he contemplated the question. "When you meet the patients," he replied carefully, "you are driven by compassion to want to do everything you can." On the other hand, researchers must avoid becoming committed to "bad models of how the brain works, whether they are Freudian or some simple pharmacologic reductionism."

Freud as Auteur

If one cannot distinguish between scientific "stories" on an empirical basis, one is more likely to be swayed by aesthetic factors. Freud's rhetorical talents have been acknowledged by friends and foes alike. In 1985 the psychologist Hans Eysenck, a bitter opponent of psychoanalysis, called Freud "a genius not of science but of propaganda, not of rigorous proof but of persuasion, not of design of experiments but of literary art. His place is not, as he claimed, with Copernicus and Darwin but with Hans Christian Andersen and the Brothers Grimm, tellers of fairy tales." The molecular biologist and neuroscientist Francis Crick, co-discoverer of the double helix, concurred. "By modern standards," Crick once wrote, "Freud can hardly be regarded as a scientist but rather as a physician who had many novel ideas and who wrote persuasively and unusually well."

Freud is not just another good writer, according to the literary critic Harold Bloom of Yale University. In *The Western Canon*, Bloom included Freud in his list of the twenty-six most important authors of all time. Freud was a key writer of the modern "Chaotic Age," along with Marcel Proust, James Joyce, and Franz Kafka.

Bloom had no illusions about the therapeutic value of psycho-analysis, which he called a form of "shamanism" that is "dying, perhaps already dead." But Freud's work, Bloom continued, "which is the description of the totality of human nature, far tran-scends the faded Freudian therapy. If there is an essence of Freud, it must be found in his vision of civil war within the psyche."

Paul Gray, the literary critic for *Time*, arrived at much the same conclusion in his famous cover story, "Is Freud Dead?" Gray ac-cepted the claims of Crews and other critics that Freud was deeply flawed, as both a scientist and a man. Gray nonetheless predicted that Freud would still endure as a literary figure. "For all of his log rolling and influence peddling," Gray wrote, "his running roughshod over colleagues and patients alike, for all the sins of commission that critics past and present lay on his couch, he still managed to create an intellectual edifice that *feels* closer to the ex-perience of living, and therefore hurting, than any other system currently in play."

Even Frederick Crews once fell under Freud's spell. Crews be-gan reading Freud as a graduate student at Princeton in the 1950s, and he was deterred from becoming an analyst himself only by the high cost of training. Early in his career, he touted psycho-analysis as a tool for literary analysis in his classes and his writings. In *The Sins of the Fathers: Hawthorne's Psychological Themes*, published in 1966, Crews argued that *The Scarlet Letter* and other works of Nathaniel Hawthorne "anticipate the findings of psychoanalysis, and the innermost concerns of his art are invariably those to which Freud attributed prime importance." Crews's book is still considered a classic of Freudian literary criticism.

The consensus seems to be that Freud achieved literary, if not scientific, truth. A number of prominent scientists have suggested that, given the record thus far, deep insights into the human psy-che may *always* be more literary than scientific in nature. Clifford Geertz, an anthropologist at the Institute for Advanced Study in Princeton, New Jersey, has proposed that debates about human nature—unlike questions in nuclear physics or molecular biology or "harder" fields of science—cannot be unambiguously resolved through an appeal to empirical evidence. Geertz has argued that his own field of anthropology is and may always be a half-literary, half-scientific enterprise. Geertz's term *faction*, which he defines as "imaginative writing about real people in real places at real times," aptly describes Freud's case histories. (Of course, if Freud was fabricating large portions of these case histories, they might not qualify even as faction.)

Similar views have been set forth by Howard Gardner, a psychologist and professor of education at Harvard University. Gardner is a codirector of Harvard Project Zero, which pursues research on the psychological underpinnings of the arts, creativity, and learning. He is best known for proposing in his 1983 book, *Frames of Mind*, and elsewhere that humans possess not a single, all-purpose intelligence but "multiple intelligences" suited to different tasks. In *Extraordinary Minds*, Gardner presented case studies of four archetypal geniuses. Gandhi was an "Influencer," someone with uncanny leadership ability; Virginia Woolf was an "Introspector," who excelled at laying bare her own psyche; Mozart was a "Master," who dominated a preexisting realm of creativity; Freud was a "Maker," who created a new intellectual realm virtually from scratch. Freud was not a bad Influencer or Introspector either, Gardner remarked.

I had expected before meeting Gardner that he would be an earnest, gentle man. His writings about multiple intelligence exuded these qualities, along with what struck me as an old-fashioned, romantic liberalism. His career seemed dedicated to the principle that education should treat each person as a unique individual with his or her own special traits that could not be captured in a single IQ score. In person, however, Gardner was as sharp-tongued as Frederick Crews. Gardner's chief target was not psychoanalysis but the entire field of psychology.

Psychology, Gardner informed me, is not only dying but already dead. He had stopped sending his membership dues to the American Psychological Association. "I thought it was a good symbolic thing to do," he said. "But I still introduce myself as a psychologist." Asked to cite the greatest achievements of psychology, Gardner replied that psychologists had learned how "not to fool themselves" into drawing unwarranted conclusions. "Conceptually and methodologically, psychology has achieved something," he elaborated. "But if you say, what are the permanent truths that psychology has amassed over the past 100 years, I don't think there are that many."

Cognitive science, neuroscience, and other mind-related disciplines could all make important contributions to our knowledge of the mind, Gardner acknowledged. The field of social psychology would continue to speculate about human culture, coining concepts and terms—such as *identity crisis* or *conventional wisdom* or *learned helplessness*—that are not rigorous scientific theories but can "help us think." Applied psychologists would continue devising IQ tests and other tools that help companies decide whom to hire or fire. But none of these fields had come close to solving the riddle of the human mind.

Gardner had first offered this critique in a 1992 lecture, "Scientific Psychology: Should We Bury It or Praise It?" He recalled that William James, although generally a scientific optimist, had occasionally expressed concern about psychology's "ante-scientific condition" and "confused and imperfect state." James once complained that "there is no such thing as a science of psychology." James's concerns about his field "have proved all too justified," Gardner declared in his lecture. "Psychology has *not* added up to an integrated science, and it is unlikely ever to achieve that goal."

Gardner granted that certain approaches to the mind, notably neuroscience, had progressed and would continue to do so. He remarked that "the years at the close of our century can well be described as the coming-of-age of brain- or neuroscience. At every level of the nervous system, from the individual synapse to the blood-flow patterns through the entire cortex, our knowledge is continuing to accumulate at a phenomenal rate." But "the phenomena of sensation, perception, or other psychological states will never be reducible" to a strictly neural theory, Gardner added.

Gardner was surprisingly critical of cognitive science, a field that he had helped to popularize with his 1985 book, *The Mind's New Science*. Cognitive scientists view the mind as an information-processing device, a computer made of flesh and blood rather than of silicon. As behaviorism declined in prestige in the late 1950s, cognitive science seemed poised to become the dominant paradigm in psychology. But cognitive science's treatment of perception, memory, attention, and reasoning has not turned out to be an obvious improvement on older psychological approaches. Moreover, Gardner added, cognitive scientists "may shortchange those aspects of reasoning or problem-solving which are characteristic of humans rather than mechanical objects."

Gardner's main complaint was that strictly scientific approaches to the mind have not advanced our understanding of psychology's core topics: consciousness, the self, free will, and personality. These subjects "seem particularly resistant to decomposition, elementarism, or other forms of reductionism," Gardner said. He contended that psychologists may advance by adopting a more "literary" style of investigation and discourse. After all, Shakespeare and Dostoevsky can tell us as much or more about the self than psychologists can. And some of the greatest figures in psychology have been those with the greatest literary skills and knowledge, such as William James.

During our conversation, Gardner pointed out that Freud too was a master of the type of literary psychology required for addressing the deepest mysteries of the mind. Freud often drew on literature and mythology in his writings, and he possessed a profoundly intuitive sense about humans. All the "great" psychologists Gardner had known possessed this intuitive capacity. "By the way, 95 percent—you can quote me on this—95 percent of psychologists" are not deeply intuitive about others, Gardner added. "They come to psychology out of chemistry, because they weren't good enough in chemistry."

Freud's Redeeming Pessimism

Psychoanalysis is a profoundly flawed paradigm; we should be thankful that it has fallen from its once-dominant position. On the other hand, the cult of Freud poses less of a danger now than the cult of Prozac, and DNA, and Darwin, and the Computer. Neo-Freudians, by challenging the hegemony of these supposedly new-and-improved paradigms, may still serve a useful purpose. "At its best," the British psychoanalyst Adam Phillips wrote in 1998, psychoanalysis "shows us both the limits of our much vaunted understanding, and what we use our so-called understanding to do."

To my mind, one of Freud's most redeeming features was his willingness to acknowledge the limits of science, including his own contributions. In *Analysis Terminable and Interminable*, published in 1927, Freud wrote, "It almost looks as if analysis were the third of those 'impossible' professions in which one can be sure beforehand of achieving unsatisfying results. The other two, which have been known much longer, are education and government." In 1933 he sounded even more sardonic: "I do not think our cures can compete with those of Lourdes. There are so many more people who believe in the miracles of the Blessed Virgin than in the existence of the unconscious."

Freud's concerns were well taken. More than a half-century after he made that statement, there is still no conclusive evidence that psychoanalysis is superior to faith healing as a mode of therapy. On the other hand, there is no conclusive evidence that any of the hundreds of alternative talking cures, based on alternative theories of human nature that have sprung up over the last century, work any better.

3

PSYCHOTHERAPY AND THE DODO HYPOTHESIS

. . . .

We've Had a Hundred Years of Psychotherapy and the World's Getting Worse.

—TITLE OF A 1993 BOOK BY THE PSYCHOANALYST
JAMES HILLMAN AND THE POET AND JOURNALIST
MICHAEL VENTURA. THE BOOK ADVOCATES A RE-
VAMPING OF PSYCHOLOGY AND PSYCHOTHERAPY
BASED ON JUNGIAN PRINCIPLES.

One recent summer day my home-town library held a sale of donated books. Wandering among tables heaped high with old paperbacks and hardcovers, I came across a fossil bed of pop psychology best-sellers: *The Politics of Experience* by the British psychiatrist and poet R. D. Laing; *Beyond Freedom and Dignity* by the psychologist B. F. Skinner, a founder of behaviorism; *Identity: Youth and Crisis* by the psychoanalyst Erik Erikson; *I'm Okay—You're Okay* by Thomas Harris, a psychiatrist and popularizer of transactional analysis; and *The Primal Scream* by Arthur Janov, a clinical psychologist and inventor of primal scream therapy, whose most famous adherents were John Lennon and Yoko Ono.

These books indicate just how fractious the history of psychology has been over the twentieth century. Like a bacterium exposed to ultraviolet radiation, psychoanalysis began rapidly mutating and evolving almost immediately after its inception.

Eventually it spawned a host of similar but competing psychologies, including ones founded by such Freudian apostates as Carl Jung and Wilhelm Reich. Independent theories of human nature also sprang up, such as behaviorism, sociobiology, and cognitive science. Almost every one of these theories boasts its own psychotherapy. By the mid-1980s, consumers could choose among more than 450 different forms of talk therapy, from active analytic therapy to Zaraleya psychoenergetic technique.

Therapies can be divided into three broad categories:

- Psychodynamic therapies, which include psychoanalysis and its derivatives, view childhood experiences, particularly those of a sexual nature, as the key to adult distress.
- Behavioral therapies take a more here-and-now approach. Based on the work of Ivan Pavlov, J. B. Watson, B. F. Skinner, and other behaviorists, behavioral therapies attempt to change harmful patterns of behavior through conditioning exercises. Behavioral therapies are often lumped together with cognitive therapies, which address harmful habits of thought rather than of behavior.
- Experiential therapies have a more philosophical and even spiritual perspective than either psychodynamic or behavioral therapies. Depression and anxiety are viewed as legitimate responses to the meaninglessness of existence. Therapists try to help patients confront and overcome their despair and alienation by recognizing their own capacity for creating meaning within their lives.

Even within a single subdiscipline there can be enormous variation. In classic psychoanalysis, the analyst is supposed to avoid giving the patient advice, but many analysts, including Freud, violated this precept. In fact, there are arguably as many types of psychotherapy as there are psychotherapists—even more, given that many psychotherapists adopt different approaches for different patients.

The accepted wisdom of late has been that psychotherapy is declining as a result of managed care and the surging popularity of drugs such as Prozac. Although anecdotal evidence for the decline of psychotherapy abounds, rigorous data are harder to come by. In 1987 7.3 million Americans, or 3.1 percent of the total population, received at least one session of psychotherapy a year on an outpatient basis, according to one federal survey. The number of outpatient psychotherapy visits totaled about 80 million and cost more than $4 billion. A 1992 study counted 100 million psychotherapy sessions, but this figure included patients in hospitals and other institutions.

A federal survey focusing on physicians, including psychiatrists, found that they were providing less psychotherapy in the late 1990s than in the previous decade, but this decline may have been offset by an increase in visits to non-M.D. therapists. Even if there has been an overall decline in psychotherapy, it still represents a major form of treatment for psychological problems—and those who provide it still constitute a powerful political force. The United States alone harbors nearly 300,000 psychotherapists. That figure includes 40,000 psychiatrists, 80,000 clinical psychologists, and 120,000 social workers. Pastoral counselors, drug addiction advisers, and other eclectic practitioners add another 50,000 or so therapists to the mix. In the mid-1990s all these groups lobbied the U.S. Congress to require firms to provide the same insurance coverage for mental illness that is provided for physical ailments such as cancer and heart disease. The so-called Mental Health Parity Act (with the inevitable loopholes) went into effect in January 1998.

The psychotherapy industry represents a vast, ongoing test of the myriad psychologies that arose during the last century. One can weight the relative merits of different psychologies by measuring the effectiveness of their corresponding therapies. If one psychotherapy proves to be superior to others at treating psychological ailments, that would represent strong circumstantial evidence for its underlying theory. Like politicians, scientific theories, especially those with medical pretensions, should perhaps be judged not for what they say but for what they do. The question shifts from "Is it true?" to "Does it work?" Early advocates for quantum mechanics could only describe the results from esoteric experiments. Later, they could submit into evidence fission reactors, transistors, lasers, and thermonuclear bombs, technologies that have altered the course of history. For many physicists, whether quantum mechanics is true is almost irrelevant; it *works*.

Determining whether a given psychotherapy works is not an easy matter. Freud sought to establish the efficacy of psychoanalysis through his accounts of individual patients, including such legendary figures as the Rat Man, the Wolf Man, and Anna O. (who coined the term "talking cure" and was actually treated not by Freud but by a colleague). As narrated by Freud, these case studies confirmed the power of psychoanalytic treatment. With the help of the brilliant analyst, the patient comes to understand the psychological sources of his or her symptoms and thus achieves relief. Of course, as various critics have shown, Freud's tales often diverged rather far from reality. Furthermore, those patients of Freud and his colleagues who improved—and some never did—might have done so in spite of rather than because of their treatment.

Case histories remain a staple of modern books about mindscience, whether they are touting psychotherapy (*The Talking Cure* by Susan Vaughan), drugs (*Listening to Prozac* by Peter Kramer), or behavioral genetics (*Twins* by Lawrence Wright). Although they often make compelling reading, case histories are no longer considered reliable forms of evidence for medical treatments. Relying on individual stories alone, one can find support for any treatment, whether Jungian therapy for depression or shark cartilage for cancer.

The best way to evaluate a medical treatment is to measure its effect on a large number of people in a controlled trial. But trials of psychological treatments are complicated by factors that do not arise in trials for cancer treatments or other more straightforward therapies. Toksoz Karasu, a psychiatrist at the Albert Einstein College of Medicine and a leader in psychotherapy assessment, once noted:

Attempting to answer the deceptively simple question, "Is therapy effective?" must take into account the diversity of theoretical and clinical approaches to psychotherapy, the difficulties in describing and measuring it as a uniform practice, the range of disorders, the variety of settings, the large number of patient, therapist, and interactional variables—seemingly ad infinitum. Furthermore, therapists ... have political, economic, and narcissistic interests that divide them not only along professional roles and identities, but along ideological lines.

Diagnosing mental illness is difficult; what appears to be depression to one psychiatrist might be diagnosed by another as schizophrenia, manic depression, or just ordinary grief. Therapists disagree, to put it mildly, over how a given disorder should be defined and even over what should be considered a disorder. The American Psychiatric Association has sought to resolve this problem with the *Diagnostic and Statistical Manual of Mental Disorders* (commonly called DSM), which was first published in 1952. But the DSM, which is generated by teams of psychiatrists whose judgments are supposed to represent the consensus of the profession, has if anything highlighted rather than eliminated the subjective nature of psychiatric diagnosis. The description of schizophrenia, one reviewer of the DSM-IV (the fourth edition, published in 1994) noted, "boils down to this: a schizophrenic is a person who thinks very odd thoughts, behaves weirdly, and suffers from bizarre delusions, which suggests that the authors of the DSM-IV either don't know what schizophrenia is or suffer from poor writing skills."

The DSM-IV does not even mention hysteria or neurosis, those two staples of Freudian psychology. Nor does it include homosexuality, which was listed as a pathology in the DSM-I and DSM-II but was deleted from later editions after lobbying by gay activists and others. Nevertheless, the number of official disorders surged from 106 in the DSM-III (1980) to more than 300 in the DSM-IV. The new categories include attention-deficit disorder ("Fails to give close attention to details or makes careless mistakes"), antisocial personality disorder ("Impulsivity or failure to plan ahead"), and dissociative fugue (an overwhelming urge to "travel away from home or one's customary place of work"). The DSM-IV reflects "the growing tendency in our society to medicalize problems that are not medical, to find psychopathology where there is only pathos, and to pretend to understand phenomena by merely giving them a label and a code number," declared the 1997 book *Making Us Crazy*, a history of the DSM. Moreover, most insurance companies will pay for the treatment only of conditions listed in the DSM.

Even if diagnosis were clear-cut, disagreements would still arise over how to measure the effect, if any, of a psychological treatment. Is the therapist the best judge of a patient's condition, or the patient? And what constitutes a positive outcome? This latter question is fraught with value judgments. While most (though not all) observers would agree that suicide represents a failed outcome, what constitutes success? Over what time period? Is the goal of therapy to promote self-understanding, happiness, or both? Should therapists help patients to become more independent or to conform to conventional social standards? How long must a patient be free of adverse symptoms before he or she is considered cured, if indeed that term is viable? Does it make sense to speak in terms of a "cure" for melancholy or anxiety, given that these may be legitimate responses to existence?

On a slightly less philosophical level, there is the problem of establishing controls for a study of psychotherapy. In a controlled study, one compares the treated population to another group that is untreated but otherwise as similar as possible. Any difference between the two groups can then plausibly be attributed to the treatment. The ideal clinical trial is double-blind; one group receives the treatment being studied, and the control group receives an inert substance, which is called a placebo. Neither the subjects nor the experimenters know who is receiving the actual treatment and who is receiving the placebo. But how does one conduct a double-blind study of psychotherapy?

Psychoanalysts Assess Themselves

Freud often excoriated those who sought empirical validation more rigorous than that provided by case studies. Large-scale statistical studies, Freud asserted, "are in general uninstructive; the material worked upon is so heterogeneous that only very large numbers would show anything. It is wiser to examine one's individual experiences." Freud's resistance to large-scale studies may have reflected his suspicion that the results would not favor psychoanalysis. Toward the end of his career, he wrote, "One ought not to be surprised if it should turn out in the end that the difference between a person who has not been analyzed and the behavior of a person after he has been analyzed is not so thorough-going as we aim at making it."

Psychoanalysts often defined therapeutic success in such a way that it was impossible to measure. The journalist Janet Malcolm drew attention to this phenomenon in her 1982 book, *Psychoanalysis: The Impossible Profession.* "That the analyst is out for bigger game than simply making the patient feel good is one of psychoanalysis's oldest and most firmly held beliefs," she wrote. According to this view, relief from distress can be an *impediment* to progress. Freud himself once chided analysts who want to "make everything as pleasant as possible for the patient, so that he may feel well there and be glad to take refuge there again from the trials of life. In so doing, they make no attempt to give him more strength for facing life."

Given these factors, it is not surprising that many analysts have expressed ambivalence or outright opposition to any effort to quantify their product's efficacy. In 1948, leaders of the American Psychoanalytic Association formed a committee to gather evidence on efficacy from members, but the committee quickly dissolved after members failed to cooperate. A committee formed in 1953 did manage to collect reports on patient outcomes from thousands of analysts. When the committee members presented the information to the association leadership in 1957, they recommended that the results be kept secret. The association sent the raw data to IBM for analysis, but many of the data were apparently lost. In 1967, the association tried to salvage the project by presenting some results—which one critic described as a collection of "anodyne factoids"—before dropping the issue once again.

Nevertheless, in the 1950s psychoanalytic institutes in Chicago, New York, Boston, San Francisco, and elsewhere began compiling data on patient outcomes. The results, which involved more than six hundred patients, were reviewed in 1991 in *Journal of the American Psychoanalytic Association* by a group led by Henry Bachrach, a professor of psychiatry at the New York Medical College at Saint Vincent's Hospital. The authors concluded that from 60 to 90 percent of the patients studied had showed "significant" improvement as a result of psychoanalysis.

These studies are so flawed that they are considered worthless by observers outside the psychoanalytic community, and even by many within it. First, assessments of patients' progress were usually made by their own analysts, who would obviously be inclined to report positive outcomes. Second, there was no attempt to compare the patients with a control group that received no therapy. Finally, investigators admitted only patients thought "suitable" for psychoanalysis. "Suitable" patients tended to be more affluent, better educated, less functionally impaired, and more motivated to receive treatment than those who were "unsuitable." In other words, those who benefited most from psychoanalysis were those who needed it least. When he presented these data at the 1992 meeting of the American Association for the Advancement of Science in Chicago, Bachrach acknowledged that the studies were not ideal. But the weaknesses "were no greater than in comparable research about other forms of psychotherapy," Bachrach asserted.

The Dodo Hypothesis

Contrary to Bachrach's claim, researchers have carried out more rigorous studies of other psychotherapies. One of the lengthiest trials was the Cambridge-Somerville Delinquency Prevention Project. Begun in 1937, the study tracked the progress of more than six hundred Boston-area boys whose average initial age was ten years old and who were thought to be at risk of delinquency. One group of boys was counseled twice a month for an average of five and a half years by social workers trained in psychoanalysis or in a then-popular "humanistic-experiential" treatment invented by the American psychologist Carl Rogers. Another group of boys received no treatment.

By 1948, there was no difference in the criminal records of the two groups. The same was true in the 1950s and in 1975. But by then some interesting differences between the two groups had emerged. Those who had been treated and had subsequently become criminals were more likely to have committed more than one crime. Researchers also found a positive correlation between the length of treatment and the degree of criminal activity. The study seemed to suggest that therapy, far from helping youths avoid a life of crime, actually increased the risk. " 'More' was 'worse,'" a researcher grimly concluded.

Discouraging results also emerged from research by Hans Eysenck, a German-born psychologist who spent most of his career at London University's Institute of Psychiatry. In 1952, Eysenck reported on a study of psychoanalysis and other "eclectic" psychotherapies. He stated that 44 percent of those undergoing psychoanalysis improved, compared to 64 percent of those receiving other therapies. But two-thirds of a group of neurotic patients showed improvement after a two-year period during which they received no treatment at all. Eysenck concluded that psychotherapy at best had no effect; in the case of psychoanalysis, it had an *adverse* effect.

Eysenck's overt hostility toward Freud—and his advocacy of alternative theories of human nature—led to charges that his conclusions were biased. He once complained that Freud's model of the mind was "too absurd to deserve scientific status." It was not a theory but "a medieval morality play" whose characters consisted of such "mythological figures as the ego, the id and the superego."

Eysenck favored genetic theories of temperament and intelligence; he was an early advocate of the notion that the relatively low IQ scores of blacks had a genetic basis.

Over the next few decades, researchers attempting to assess psychoanalysis and other psychotherapies arrived at a more positive conclusion than Eysenck had: people receiving psychotherapy generally improved in comparison to control subjects receiving no therapy. (Typically the controls would be placed on a "waiting list" for therapy.) In the late 1970s the U.S. Congress, considering legislation that would require insurers to cover psychotherapy, directed the now-defunct Office of Technology Assessment to assess psychotherapy. "Although the evidence is not entirely convincing," the Office of Technology Assessment reported back to its congressional overseers in 1980, "the current literature contains a number of good-quality research studies which find positive outcomes for psychotherapy."

One of the most influential reports on psychotherapy was "Comparative Studies of Psychotherapy: Is It True That 'Everybody Has Won and All Must Have Prizes'?" published in *Archives of General Psychiatry* in 1975. In the paper Lester Luborsky, a professor of psychology at the University of Pennsylvania, and two colleagues reviewed studies of different therapies. They concluded that psychotherapy was worthwhile; those who received therapy generally fared better than those who did not. On the other hand, none of the therapies stood out; all were roughly as effective as each other. To dramatize the import of this finding, Luborsky and his colleagues recalled an episode in *Alice's Adventures in Wonderland* in which Alice and other characters adrift in a sea of her tears washed up onto an island, soaking wet. There they encountered a Dodo bird who recommended that they dry off by running a race around the island. The scene unfolded as follows:

First [the Dodo] marked out a race course, in a sort of circle ("the exact shape doesn't matter," it said) and then all the party were placed along the course, here and there. There was no "One, two three, and away!" but they began running when they liked, so that it was not easy to know when the race was over. However, when they had been running half an hour or so, and were quite dry again, the Dodo suddenly called out, "The race is over!" and they all crowded round it, panting, and asking, "But who has won?"

This question the Dodo could not answer without a great deal of thought, and it sat for a long time with one finger pressed upon its forehead (the position in which you usually see Shakespeare, in the pictures of him), while the rest waited in silence. At last the Dodo said, "*Everybody* has won, and all must have prizes!"

Luborsky credited the psychologist Saul Rosenzweig with coining the phrase "Dodo hypothesis" in a 1936 paper. Rosenzweig had postulated that all psychotherapies might be equally effective but had offered no evidence for his assertion. Luborsky has continued to update his findings over the years, and he has become even more convinced that the Dodo hypothesis is correct. Studies indicating the superiority of a particular therapy usually reflect what Luborsky calls an "allegiance effect"—the tendency of researchers to find evidence for the therapy that they practice or favor.

On the other hand, Luborsky has allegiances of his own. When I called him to discuss his findings, he took pains to reject the claim of Hans Eysenck and others that psychotherapy confers no benefits. "There is a huge amount of evidence that psychotherapy works," Luborsky told me. I asked if this evidence might stem from what could be called a meta-allegiance effect; after all, a researcher who favors a specific therapy also presumably believes in the benefits of talk therapy in general. Luborsky said he had not examined—or even considered—that possibility, but he doubted whether it was true.

Of all the talking cures, moreover, Luborsky is clearly committed to psychodynamic therapy, which is a kind of psychoanalysis lite. Luborsky revealed this allegiance in a chapter that he cowrote for *Psychodynamic Treatment Research: A Handbook for Clinical Practice*, published in 1993. The main topic of the chapter was the Dodo hypothesis. Toward the end of the chapter, Luborsky and his colleagues attempted to answer some questions raised by their review of the Dodo hypothesis with an "imaginary dialogue." One exchange in the dialogue went as follows:

QUESTION: Don't you feel, despite all the evidence for the non-significant difference effect, that [psychodynamic] therapies have some special virtues to offer that are still not recognized?

ANSWER: I'm glad you asked that. The answer is definitely yes. The studies have not yet dealt with possible long-term benefits. Nor have they dealt well enough with the distinction between changes in symptoms and changes in general adjustment.

Luborsky, who was trained as a psychoanalyst, acknowledged to me that he remains devoted to psychoanalytic theories of the mind. He hoped that further research would identify the beneficial components of psychodynamic therapy so that it could be made more effective. Luborsky has advanced an elaborate hypothesis, which draws heavily on the Freudian concept of transference, about the relationship that develops between patients and therapists in successful therapy. The essence of this hypothesis, he told me, is that "an alliance is formed, new insights are gained. That's the simplest way to say it."

Psychotherapy as Placebo

In 1993 Luborsky divulged his continuing allegiance to psycho-analysis in a letter responding to Frederick Crews's article "The Unknown Freud" in the *New York Review of Books.* Luborsky rejected Crews's assertion that psychoanalysis is "an indifferently successful and vastly inefficient method of removing neurotic symptoms." Psychoanalysis is "probably at least as good as other forms of psychotherapy," Luborsky insisted. "The evidence for this is that for all treatment comparisons involving different forms of psychotherapy the overwhelming trend is for non-significant differences in benefits." In his retaliatory blast, Crews asked why, if all psychotherapies are equivalent, anyone in his right mind would choose the most expensive and time-consuming psychotherapy of all, psychoanalysis. "The real import of Luborsky's work," Crews added, "is that psychotherapies succeed (when they do) thanks to factors that they all share—that is, placebo factors."

The placebo effect has long been a bugaboo of modern medicine. The term *placebo* derives from the Latin term for "I shall please," the first phrase in the Catholic vespers for the dead. The vespers themselves were sometimes called *placebos*, as were professional mourners hired to sing the vespers. The term came to refer to sycophants and flatterers and, over time, to sham treatments that physicians used to placate patients.

Arthur Shapiro, a professor of psychiatry at the Mount Sinai School of Medicine in New York, was an authority on the placebo effect. In an essay that he cowrote just before his death in 1995, he noted that "until recently, the history of medical treatment is essentially the history of the placebo effect." An examination of all the thousands of remedies that predated modern medicine suggested that "with only a few possible but unlikely exceptions, all were placebos." One popular cure-all, called theriac, consisted of dozens of ingredients, including the skin of a snake (which was thought to have restorative powers). First described by Galen almost eighteen hundred years ago, theriac was prescribed by European physicians through the end of the nineteenth century.

Healers realized long ago that the most important ingredient of these remedies might be patients' belief in them. Galen once wrote: "He cures most successfully in whom the people have the most confidence." It has long been a tenet of medical lore, Arthur Shapiro noted, that patients respond better to new drugs than to older, more established ones. This syndrome is summed up in an old doctor's dictum: "Hurry, hurry, use the new drug before it stops healing." The introduction of a novel drug often generates unrealistically high expectations on the part of both patients and physicians—expectations that can become self-fulfilling. Over time, as the drug's novelty fades and its side effects and limitations become more apparent, it becomes less effective.

Far from being illusory or purely psychological, the placebo effect can bring about striking physiological benefits. A landmark study published in 1955 by Henry Beecher revealed that placebos can provide measurable relief from ailments such as asthma, high blood pressure, and warts. The percentage of patients responding to a placebo for various ailments ranged from 30 to 40 percent, according to Beecher. The effects of the placebo sometimes "exceed those attributable to potent pharmacologic action," he stated.

The effects are much higher for some procedures. In the 1950s researchers conducted a clinical trial of arterial ligation, a surgical treatment for chest pain caused by insufficient blood flow to the heart. Patients who underwent the procedure were compared to a control group who received chest incisions but no further surgery. Seventy-six percent of the patients receiving arterial ligation improved. That result would have seemed quite good except for the fact that *100 percent* of those receiving simple incisions—the placebo procedure—responded favorably. (Arterial ligations have been abandoned.)

The flip side of the placebo effect is the nocebo effect, in which negative expectations become self-fulfilling. Incredibly, even non-human animals are subject to these supposedly "psychological" effects, according to experiments performed in the 1970s. The psychologist Robert Ader of the University of Rochester gave rats saccharine-flavored water containing cyclophosphamide, a compound that causes nausea and severely suppresses the immune system. All the rats became sick, and many died. Ader then gave the survivors saccharine-sweetened water containing no cyclophosphamide. These rats, conditioned by their previous exposure, became sick as well, and some even died. The sweet-tasting water alone was enough to suppress their immune systems fatally.

The psychiatrist Arthur Shapiro argued that the "power of the placebo" is reflected in the ubiquity of fraud (quackery, $30 billion annually), faith (religious and psychic healing), fallacies (vitamins, organic diets, excessive jogging, holistic treatments, and alternative treatment, $13.9 billion annually), and fads (new age, lifestyle change or self-help modalities, such as ecotherapy or communing with nature and meditative immunotherapy, to encourage the growth and strength of good white blood cells and destroy malignant cells). Despite the reasonable expectation that the use of these therapies would decrease as scientific knowledge increased, these therapies continue to appear, disappear, and reappear in slightly different guises, as in a witch's bubbling cauldron.

The placebo effect might also be the primary ingredient underlying psychotherapy, Shapiro suggested. He compared psychotherapy to the medieval potion theriac, an "unsystematic myriad of nonspecific elements mixed together in the hope that some will be effective." Shapiro added: "Although there is general agreement that psychotherapy is useful, beneficial, and effective for many patients, as is true of many notable placebo treatments, the knotty question remains—is psychotherapy more than placebo?"

Perhaps not, according to Jerome Frank, a former professor of psychiatry at the Johns Hopkins University School of Medicine. Frank reluctantly reached this conclusion as a result of his own research. In the late 1950s, he and his colleagues provided depressed patients with one of three forms of therapy: weekly individual therapy, weekly group therapy, and minimal individual therapy, which consisted of just one half-hour session every two weeks. "To our astonishment and chagrin, patients in all three conditions showed the same average relief of symptoms," Frank recalled.

On the basis of this and other investigations, Frank concluded that "relief of anxiety and depression in psychiatric outpatients by psychotherapy closely resembles the placebo response, suggesting that the same factors may be involved." The specific theoretical framework within which therapists work has little or nothing to do with their ability to "heal" patients, Frank contended. The most important factor is the therapist's ability to persuade patients that they will improve.

"I do think my views have been borne out," Frank told me when I called him in 1996. He doubted whether science can demonstrate the efficacy—or lack thereof—of any particular psychotherapy, since science cannot pinpoint or measure the qualities that enable a particular therapist to induce the placebo effect in a particular patient. Given this perspective, Frank was bemused by the vehemence of the attacks on Freud. "People have been attacking Freud because he wasn't a scientist, but that misses the point. He was a great mythmaker."

Frank spelled out his views in *Persuasion and Healing*, originally published in 1961 and reissued many times since then (most recently in an edition co-written with his daughter, Julia Frank, also a psychiatrist). Presaging Howard Gardner's views of psychology, Frank suggested that psychotherapy should be considered as a branch not of science but of rhetoric, the art of persuasion: "The methods of both psychotherapist and rhetorician purport only to approach, not to achieve, truth. That is, the truths of these disciplines are probable, not certain." Frank also likened psychotherapy to literary criticism. A patient's clinical history "resembles a text, and psychotherapy a collaborative effort of patient and therapist to discern its meaning." There is no single, correct way to treat a patient, just as there is no single, correct way to read a book. Different readers will find different meanings in the same text, and different psychotherapists will interpret the same patient's remarks and clinical history in different ways.

Of course, different patients respond to different therapists. One will respond to a therapist who is warm and empathetic, another to a therapist who is authoritative and cool. For some patients, "faith in science still seems to provide the predominant source of symbolic healing power," Frank said. Psychotherapists may thus enhance their effectiveness by displaying scientific "symbols of science," such as a medical school diploma or a physician's jacket or a stethoscope. But many people, Frank noted, prefer healers who espouse nonscientific mythologies. For these patients, shamans and faith healers may provide more relief than psychiatrists or clinical psychologists.

Here-and-Now Therapies

Many experts continue to chafe at the notion that all psychotherapies are more or less equal in their ability to help patients. The Freud basher Frederick Crews insisted to me that psychoanalysis is worthless at best, and recovered-memory therapy can be devastatingly harmful. Crews favored treatments such as cognitive therapy, which takes a patient's problems at face value and promotes a "here-and-now, helpful attitude." If the patient complains that he has a lousy job, or a bad sex life, or difficulty making friends, the therapist should "take it literally, and find out where this person is strong, find out what resources this person has, and help the person cope."

Crews is hardly alone in favoring cognitive therapy. Martin Seligman of the University of Pennsylvania, a former president of the American Psychological Association and a well-known outcome researcher, has argued that recent studies give an edge to cognitive therapy. The American Psychological Association has recommended cognitive therapy for panic disorder and other problems. One reason for the popularity of cognitive therapy—and a closely related variant, cognitive-behavioral therapy—is its specificity. The patient and therapist focus on precisely those symptoms or maladies that are bothering the patient—smoking, drinking, irritability toward a spouse or children, obsessive handwashing, bulimia. This pragmatic, focused approach appeals to many patients and to virtually all managed care providers. (In one variant of cognitive therapy, the therapist repeatedly orders patients to describe anxiety-provoking situations and repeatedly interrupts them by screaming, "Stop!")

In 1996 the *New York Times* health columnist Jane Brody reported that cognitive therapy can lead to long-term improvement after relatively short periods. Brody led off her column with the oblig-atory anecdote about a middle-aged woman who "had grappled with the crippling effects of depression almost her entire life. . . . Years of psychotherapy, including analysis, did nothing to ease her psychic pain—nothing, that is, until she began seeing a cognitive therapist. 'He saved my life,'" Brody quoted the woman as saying. Brody went on to assert that "studies have shown that the results of cognitive therapy are long-lasting, with relapse rates far lower than with other modes of treatment."

According to Lester Luborsky, author of the 1975 Dodo hypoth-esis paper, trials giving the edge to cognitive therapy or its variant, cognitive-behavioral therapy, are invariably tainted by the alle-giance effect. Luborsky's skeptical assessment was corroborated by a study of cognitive-behavioral therapy undertaken by M. Katherine Shear, a psychiatrist at the University of Pittsburgh, and three colleagues. The researchers examined patients suffer-ing from panic disorder, a condition marked by the sudden onset of extreme, unwarranted fear. Cognitive-behavioral therapy is thought to be particularly well suited for treating panic disorder.

The researchers divided their subjects into two groups. One group received twelve sessions of standard cognitive-behavioral therapy, which entailed mental and physical exercises designed to counteract panic. The other group received a kind of placebo therapy called "reflective listening," in which the therapists lis-tened sympathetically to the patients but provided them with no specific guidance. Both sets of patients responded equally well. These data, Shear and her colleagues concluded in the May 1994 *Archives of General Psychiatry*, "raise questions about the specificity of cognitive behavioral treatment."

The Myth of Expertise

The Dodo hypothesis has at least two important corollaries, which were revealed in a 1977 article in *American Psychologist* by the psychologists Mary Smith and Gene Glass. After analyzing 375

studies of psychotherapy, Smith and Glass arrived at several conclusions. They found, as Luborsky's group had, that psychotherapy
works: treated patients fared better than the untreated. Smith and
Glass also corroborated the Dodo hypothesis: all the therapies
produced roughly equivalent outcomes.

Psychotherapists could live with these findings, but they were
less prepared for two others reported by Smith and Glass. First,
there was no correlation between the amount of time spent in
therapy and the benefits it conferred. Second, there was no correlation between the effectiveness of the therapists and their cre-
dentials or experience. In other words, psychiatrists, who have
graduated from medical school; psychologists, who have a Ph.D.;
and social workers, who have a master's degree, are all equally effective, or ineffective. Nor does the ability of therapists to help patients improve with experience.

Other investigators tried to refute the findings of Smith and
Glass, in vain. In one experiment, patients complaining of such
ailments as anxiety and depression were randomly assigned to
two different groups of "therapists": one group consisted of
genuine, professional psychologists and the other of university
professors with no background in psychology. The patients responded as well to the pseudotherapists as to the real ones. The researchers who conducted this experiment were not thrilled with
their results. "Professional psychologists," they insisted, "by virtue
of their training and clinical experience, are clearly much better
equipped to deal with the vagaries and vicissitudes encountered
in interactions with most patients." Given the researchers' own
data, this sentiment was no more than wishful thinking.

Another observer who initially found the results of Smith and
Glass hard to believe was Robyn Dawes, a professor in the department of social and decision sciences at Carnegie-Mellon University. In the late 1970s, Dawes (then at the University of Oregon) and
a colleague reanalyzed the studies examined by Smith and Glass
and arrived at the same conclusion. This exercise spurred Dawes
to become a critic of guilds such as the American Psychological
Association and the American Psychiatric Association. Both these

organizations have argued that the proliferation of psychotherapists who lack medical degrees or doctorates will harm patients.

But as Dawes pointed out in his 1994 book, *House of Cards: Psychology and Psychotherapy Built on Myth,* "Those claiming to be mental-health experts—including many psychiatrists—often assert that their 'experience' allows them to apply principles of psychology in a better manner than others could, but the research evidence is that a minimally trained person applying these principles automatically does at least as well." He urged that "in attempting to alleviate psychological suffering, we should rely much more than we do on scientifically sound, community-based programs and on 'para-professionals.'"

Dawes challenged the notion that clinical psychologists possess special knowledge and methods that enable them to discern a patient's past and predict his future more accurately than laypeople can. One well-worn technique for assessing personalities is the Rorschach test, which requires a subject to disclose thoughts or feelings evoked by an ink blot. A Rorschach test figured prominently in the 1958 melodrama *I Want to Live!* which starred Susan Hayward as an accused murderer. After giving Hayward a Rorschach test, a psychologist played by Theodore Bikel announced that she must be innocent. (Hayward was still sent to the gas chamber.)

There is no evidence that Rorschach tests or similar techniques employed by psychologists when they interview patients have any special diagnostic or predictive power, Dawes asserted. So-called actuarial methods have proved to be much more reliable for predicting behavior than the clinical methods of psychologists, according to Dawes. Actuarial methods, which are employed by insurance companies and other industries, predict the behavior of an individual based on the record of other people who are demographically similar. For example, an insurance company predicts the accident risk for a twenty-five-year-old, unmarried male with two speeding convictions based on the accident record of males whose age, marital status, and driving record are identical. Actuarial methods have consistently proved to be superior to so-called clinical methods in predicting the future behavior of psychiatric patients and criminals, Dawes said.

Touring the Psychiatric Museum

Not surprisingly, some mental health experts have concluded, based on the record of psychotherapy to date, that everybody has *lost*, and *none* must have prizes. In December 1995 the chairman of psychology at Rutgers University sent members of his department a memo stating:

> It is not clear to me why the University should be involved at all with a professional psychotherapy-related school. The value of advanced training for talk therapy has been seriously questioned by recent data which show quite conclusively that such training is totally unnecessary (see Robyn Dawes's *House of Cards*).

The chairman recommended that Rutgers abolish its graduate school of applied and professional psychology. (As of 1999, the graduate program had not been terminated.) Another advocate of such steps is the psychiatrist E. Fuller Torrey. In his 1992 book *Freudian Fraud*, Torrey complained that psychoanalysis and other psychotherapies were diverting precious resources away from those who truly needed them, the severely mentally ill. He remarked that the vast majority of the nation's psychiatrists, psychologists, and social workers "spend their time doing counseling and psychotherapy based, directly or indirectly, on Freudian theory." People with serious mental illnesses such as schizophrenia and manic depression "require medication and rehabilitation rather than discussions about early childhood experiences," Torrey asserted. Torrey, whose sister is schizophrenic, deplored "the sad spectacle of approximately 200,000 untreated mentally ill individuals among the nation's homeless population, despite the fact that America has more mental health professionals than any other country; the scene is one more legacy of Freudian theory."

I gained a somewhat different perspective on the problem of the severely mentally ill after I toured a museum maintained by the Hudson River Psychiatric Center, just north of Poughkeepsie, New York. Upon hearing about the museum in the fall of 1997, I called the center and was put through to Roger Christenfeld, a clinical psychologist who served as director of research. He said that the museum could be seen only by appointment; he would be happy to give me a tour. Giving me directions to the center, he told me to keep an eye out for signs saying, "Abandon all hope ye who enter here." His tone was so flat that it took me a moment to realize he had been joking.

The main building of the Hudson River Psychiatric Center was just as Christenfeld had described it: a "classic, red-brick, Victorian-Gothic asylum." It looked both forbidding and forlorn; weathered plywood boards sealed the windows of two wings that had once housed patients but were now unoccupied. The asylum stood atop a long, gently sloping lawn that served as a small golf course. Although it was a rainy, blustery day, a man and woman in windbreakers and hats were hitting golf balls toward a flag. Patients or employees? I wondered. (Definitely employees, I learned later; inpatients cannot roam the grounds.)

Walking through an unpainted picket fence, I entered the administration building and made my way to Christenfeld's office. A middle-aged man of medium build wearing a sports jacket and slacks, he exuded an odd formality; when he spoke, he put the heels of his shoes together and stood at awkward attention, as if he were posing for a painting. But his fastidious, even pedantic manner concealed a sardonic sense of humor.

As we made our way out of the administration building to the museum, Christenfeld offered me his rather jaded perspective of the mental health profession. He informed me that some mental health advocates, rejecting the term *patient* as demeaning, have sought to replace it with other terms. "It's undergone several changes. It's hard to keep up. After patients they became *clients*. Af-

ter clients they became—one might think this might be the ultimate American appellation—they became *consumers*, consumers
of mental health services. And we established even a treatment
mall, where you could come and interact with the treatment
providers. And last I heard, various interest groups, they want to be
called *survivors* of mental health care, or survivors of the mental
health system, which they regard sometimes in an adversarial
way."

Those who treat the mentally ill are equally faddish, Christenfeld said. "At any given moment, psychiatrists are more or less
united in their ideology and their belief, particularly in the etiol-
ogy of major mental disorders," he said. "Right now, of course,
most psychiatrists believe that the major mental disorders are biochemical, physiological, genetic. Now, I've been in this business
long enough to recognize that it's cyclical. When I got into it,
everybody believed that it had to do with disorganization and bad
child rearing and social stress, but if we just fixed up how people
brought up their kids and how they lived, it could make a real difference. There is some indication that these beliefs co-vary in
time with the political climate of the culture."

Christenfeld was not a fan of psychoanalysis. It is "bereft of
proof," he said. "It is essentially un-evaluable, and every time you
try [to evaluate it] it shows that it doesn't really have any measurable impact." Belief in psychoanalysis is akin to belief in the trans-
mutation of souls, he continued, "which is why in a managed care
environment it has become a rather fringe, boutique enterprise."

But if psychoanalysis is a "quasi-religion," Christenfeld said, so is
the biologically oriented psychiatry that has replaced it. It is not
surprising that modern psychiatrists had become so biologically
oriented, Christenfeld said. "Psychiatrists have always suffered in
the medical profession," he elaborated. Particularly during the
heyday of psychoanalysis, psychiatrists were seen as physicians
who didn't operate on patients or give them drugs but "just sat
there and talked." To distance themselves from this image, many
psychiatrists went to the opposite extreme of viewing mental illness as a strictly physiological disorder that requires strictly phys-

iological treatments. They insist that "it's all there in your neuro-transmitters and your receptors. We don't want to hear about all of this attitude business." But this approach had not been much more successful than psychoanalysis, Christenfeld said.

Christenfeld favored the view that schizophrenia stems from both genetic and experiential factors. "The genetic component simply sets your threshold. Some people, with only a small amount of stress, will become schizophrenic—a parking ticket, a failed calculus exam." Others "have a very high threshold, and will require three months of frontline combat before they go to the edge. Essentially, it's a theory that everybody has a breaking point."

Finally we arrived at the museum, which resembles an over-sized, decrepit mausoleum. As we stepped inside, Christenfeld announced: "Here, in any case, in objective form, is the history of psychiatry, at least since the founding of this hospital in 1871." It was a single, cavernous room, dimly lit. Paint chips that had fallen from the ceiling flecked the varnished wood floor. Just inside the door were several ancient restraining devices. One was a high-backed wooden chair with straps at the base and on the arms and a hole in the seat. At head level was a screen-covered box, which prevented patients from biting or spitting on others. This "Rush tranquility chair" had been designed in the early 1800s by Benjamin Rush, who is often called the Father of American Psychiatry.

Beside the tranquility chair was the "Utica crib," which looked like a coffin with bars across the top and sides. The crib would be suspended from chains and swung back and forth, ostensibly to soothe the prisoner. "But you can imagine pretty much what it would feel like to actually *be* in there," Christenfeld said, staring at the crib. Hanging from the wall behind this crib was a straitjacket, which, Christenfeld told me, is now rarely used in mental hospitals. Today patients being admitted to a hospital can indicate which form of restraint they prefer if or when they become uncontrollable. The most common physical restraints are handcuffs that clip to a belt around the patient's waist and straps that pin a patient to a bed.

Strolling around the museum, we encountered somewhat more benign artifacts from the early twentieth century: an ancient desk and a rocking chair; a large drum with "HRSH [Hudson River State Hospital] Band" printed on it; a cobbler's bench covered with tools and bits of leather; a female mannikin wearing a blue dress and white apron, the uniform of students at a nursing school formerly attached to the asylum. Dusty glass cases contained bolts of cloth, bottles, fans, combs, and (to my surprise) straight-edge razors. The hospital formed a self-sufficient "total community," Christenfeld said. Patients ran a farm, raised and slaughtered livestock, made their own clothes, shoes, and furniture.

The ostensible purpose of the hospital was to expose patients, most of them from New York City, to "beautiful views and country air" so that they could get better and return to society. But once patients were admitted to the hospital, they rarely left. The philosophy of mental hospitals of that era was "out of mind, out of sight," Christenfeld said. "People go out of their minds, get them out of your sight." Laughing at the aphorism, I asked if it had been passed down through generations of psychiatric workers. Actually, Christenfeld said, he had invented it. "I don't require attribution," he murmured.

The population of the Hudson River Psychiatric Center peaked in the 1950s at about 6,000 patients, Christenfeld said. During the 1960s and 1970s, like other publicly funded asylums across the country, it began discharging patients in droves. The center now served primarily as an outpatient hospital; it housed only 350 inpatients, most of them severely schizophrenic. "Nobody is here who doesn't display a *very* good reason for being here," Christenfeld said.

My tour with Christenfeld left me with mixed feelings. It was hard, seeing the antique Utica crib and Rush chair, not to think that psychiatry had made tremendous progress in the last century. The fact that the Hudson River Psychiatric Center housed so few patients compared to its heyday also seemed to represent a step forward. But just how genuine was that progress? The Utica crib and Rush chair gave way to straitjackets and padded cells and shock treatments and lobotomies. Were these better, or just different?

The advent of powerful anti-psychotic drugs in the 1950s had brought about by far the biggest change in mental health care. They had enabled many psychotic patients to leave hospitals like the Hudson River Psychiatric Center, where they were once doomed to spend their entire lives, and live instead in community-based homes or even on their own. The number of patients housed in state asylums dropped from 559,000 in 1955 to fewer than 70,000 by the mid-1990s.

But deinstitutionalization, which was promoted as a progressive, humane step in the treatment of the mentally ill when it began in the 1960s, has left a mixed legacy. Many of those who would formerly have been housed in mental hospitals live in group homes, with their families, or in their own homes. Many others are homeless, or worse. "Prisons Replace Mental Hospitals for the Nation's Mentally Ill," declared a headline in the *New York Times* on March 5, 1998. The article reported that 10 percent of the 2 million inmates in the United States are believed to suffer from severe mental illness. "Advocates for the mentally ill say the clock is being turned back to the 19th century," the article concluded, "when it was common in the United States to confine people with mental illness in jails."

E. Fuller Torrey was quite justified in expressing outrage in *Freudian Fraud* at the plight of the severely mentally ill. But his claim that Freud and other promulgators of psychotherapy are to blame for the situation is not entirely fair. Torrey himself, in a 1997 essay for the *Chronicle of Higher Education*, offered a more nuanced view of deinstitutionalization. One important factor, he acknowledged, was the belief of social activists that the mentally ill were just another oppressed minority, like blacks and women; these activists convinced lawmakers to make it much more difficult for patients to be committed to mental hospitals without their consent. But another important factor, Torrey wrote, was "the widespread hope that the new drugs would cure people." Indeed, the shift of the mentally ill from hospitals into prisons and onto streets arguably stemmed from excessive faith not in psychoanalysis and other talking cures but in a new panacea, psychopharmacology.

4

PROZAC AND OTHER PLACEBOS

. . . .

In time, I suspect we will come to discover that modern psychopharmacology has become, like Freud in his day, a whole climate of opinion under which we conduct our different lives.

—PETER KRAMER, *LISTENING TO PROZAC*

I n May 1996, the American Psychiatric Association convened in New York City for its 149th annual meeting. The contrast between this conference and the one for psychoanalysts that I had attended at the Waldorf-Astoria two months earlier was striking. The Waldorf-Astoria meeting had drawn only about four hundred participants. The American Psychiatric Association enticed some sixteen thousand psychiatrists and other workers in the mental health industry to the cavernous Jacob Javits Convention Center on New York City's grimy West Side and to the Marriott Hotel in mid-Manhattan. The lecture fare was eclectic, ranging from "Kids Who Kill" and "The Psychobiology of Binge Eating" to emerging markets for psychiatric services. One big "area of opportunity," revealed Melvin Sabshin, medical director of the American Psychiatric Association, is forensic psychiatry, which involves examining and treating mentally ill criminals. "We have more people with psychiatric disorders in jails and prisons than in hospitals," Sabshin explained. What mental health experts such as E. Fuller Torrey saw as a tragedy, Sabshin saw as an opportunity.

The buzzword was *parity*—the principle that insurance companies should provide the same coverage for mental disorders that they provide for physical ailments. A bill calling for mental health parity had won approval from the U.S. Senate in April after heavy lobbying by the American Psychiatric Association but still had to run the gauntlet of the House of Representatives. At a "town meeting" held to discuss political issues relevant to psychiatry, the audience cheered when Marge Roukema, a Republican representative from New Jersey and a fierce advocate of parity, proclaimed, "This is about fairness." The vast majority of people who see therapists, argued Roukema (who acknowledged that she was married to a psychoanalyst), are not self-absorbed neurotics like the ones depicted in Woody Allen films but people with a real need. Jay Cutler, director of government relations for the American Psychiatric Association, took the stage to urge members to employ a computerized lobbying device set up at the meeting; members could plug their names and zip codes into the computer, and it would send a form letter demanding parity to the appropriate legislators.

Other speakers voiced concern about the economic threat posed to psychiatry by psychologists and social workers, who generally charge less than psychiatrists do. On the other hand, unlike psychologists and social workers, psychiatrists can prescribe drugs, which are cheaper than protracted talk therapy and thus far more appealing to managed health care companies. The best-attended sessions at the conference were breakfasts and dinners sponsored by Pfizer, SmithKline Beecham, and other pharmaceutical manufacturers. While gobbling up omelets and chicken breasts, hundreds of psychiatrists listened to drug company representatives tout the benefits of their products for insomnia, obsessive-compulsive disorder, and depression.

The dominance of the drug companies was even more apparent in the exhibition hall, which sprawled over almost an entire floor of the Javits Center. Drug companies, publishers, and other peddlers of psychiatric wares had created a small town, complete with street signs with names such as "Norepinephrine Way" (norepinephrine is a neurotransmitter). The largest display—and the first one that a visitor encountered upon entering the exhibitors' town—belonged to Eli Lilly. At the center of Lilly's exhibit stood a golden obelisk some twenty feet high with the word *Prozac* emblazoned on its crown in iridescent red. Draped around the obelisk was a banner that proclaimed, "Knowledge Is Powerful Medicine."

Surrounding the obelisk were a dozen or so interactive televisions and humans dispensing information on this best-selling psychiatric drug of all time. The salespeople were young, attractive, impeccably groomed, clad in sleek suits and radiant smiles. Only slightly less impressive shrines to Paxil, Effexor, Zoloft, and other mood regulators stood nearby. Not everyone at the meeting embraced the better-living-through-chemistry philosophy. At a session entitled "The Future of Psychotherapy," Gene Usdin, a psychiatrist at the Ochsner Clinic in New Orleans, fretted, "At this point, I don't think the future is very good." The main reason, he said, is that "we are selling our souls" to the pharmaceutical firms. The fact that only twenty people were listening to him—compared to the hundreds who showed up for the lectures sponsored by the drug companies—lent some support to his complaint.

Another dissenter was a sales representative for Somatics, Inc., a firm based in Lake Bluff, Illinois, that had erected a modest booth in the shadow of the Prozac pavilion. "Drugs have big problems" for many patients, the Somatics salesman confided to me. His company, he claimed, manufactured equipment for a far more effective treatment for severely ill patients: electroconvulsive therapy, also known as shock treatment. Although shock treatment is still generally used only as a last resort, the salesman told me, that situation is changing; psychiatrists and patients alike are coming to recognize how effective shock therapy is, not only for ordinary depression but also for manic depression and schizophrenia. In addition to its Thymatron shock delivery system, which included such features as a "postictal suppression index" and "Chrona elapsed-time lighted display" (according to a brochure), Somatics is the first company to offer reusable mouth guards in two sizes; the smaller one is designed especially for women to "minimize risks of tooth fracture or loss."

Fevers, Comas, and Other Cures

The history of modern psychiatry can be viewed as a contest between psychological therapies, notably psychoanalysis, and physiological ones, notably drugs. Many observers see psychiatrists' growing reliance on medication and other physiological remedies as a triumph of reason over irrationality. That is the theme of *A History of Psychiatry*, by the historian Edward Shorter of the University of Toronto. Shorter made his position clear in his preface: "If there is one intellectual reality at the end of the twentieth century, it is that the biological approach to psychiatry—treating mental illness as a genetically influenced disorder of brain chemistry—has been a smashing success. Freud's ideas, which dominated the history of psychiatry for the past half century, are now vanishing like the last snows of winter." Frederick Crews had approvingly quoted from this passage in his Freud-bashing lecture at Yale.

Ironically, Shorter's own account demonstrated that biological psychiatry, far from being a "smashing success," had produced some of the most horrific treatments in the history of modern medicine. Reading Shorter's descriptions of these biological remedies, one can understand why psychoanalysis became so popular for treating not only garden variety emotional disorders but even, to a lesser extent, psychosis. As one reviewer of *A History of Psychiatry* noted, psychoanalysis "sounds relatively benign" compared to the treatments Shorter described.

Some of the early physical therapies seemed harmless enough. Freud himself employed electrotherapy, which exposed patients to low levels of electric current. (Electroconvulsive therapy, which calls for doses strong enough to produce siezures, came later.) Another nineteenth-century treatment that lasted well into the next century was hydrotherapy, also called the water cure, which involved immersing patients in very hot or very cold water, douching them, or spraying them with powerful jets.

As it became obvious that these approaches had little value, psychiatrists turned to more desperate measures. One therapy popular in the early twentieth century was the fever cure, invented by the Austrian psychiatrist Julius Wagner-Jauregg. Observing a psychotic patient briefly become lucid after contracting a severe bacterial infection, Wagner-Jauregg speculated that high fever might alleviate psychosis. Over the next few decades, he and other researchers tested this hypothesis by infecting mentally ill patients with malaria, tuberculosis, typhoid, and other diseases. In 1927 Wagner-Jauregg was awarded the Nobel prize for his research, which Edward Shorter described in *A History of Psychiatry* as "an epochal moment not just in the history of psychiatry but the entire history of medicine." Other historians have concluded that the evidence favoring fever therapy was weak at best.

The fever cure was soon displaced by insulin-coma therapy, introduced in the 1930s by the Austrian psychiatrist Manfred Sakel. Sakel was experimenting with insulin as a treatment for morphine addiction when he accidentally gave an overdose to an addict who happened to be psychotic. An overdose of insulin induces a potentially lethal coma. The psychotic patient, after emerging from his coma, seemed less disturbed, and Sakel began deliberately injecting insulin overdoses into schizophrenics. He reported extraordinary success rates: thirty-five out of fifty patients were completely cured, and nine showed partial improvement. Insulin-coma therapy quickly spread to the rest of Europe and to the United States.

Researchers experimented with other drugs that triggered not comas but convulsions. A camphor-like substance known by the trade name Metrazol induced vomiting and seizures resembling those caused by epilepsy. The drug "was never a big success," Shorter said, "mainly because it was too unreliable in producing fits and was feared by patients." Even some psychiatrists recoiled from the treatment. "The sight of the artificially produced attack of epilepsy," one recalled, "especially of the contorted blue faces, was so awful to me that I sought to get away from the room whenever I could."

The "sleep cure" knocked patients out for weeks at a time with bromide salts and other sedatives. (Oddly, antidepressant effects have also been attributed to sleep *deprivation*.) The Canadian psychiatrist Ewen Cameron developed a variant of the sleep cure that he called "depatterning" or "brainwashing." Cameron rendered patients unconscious or semiconscious for long periods with barbiturates; often he administered shock therapy as well. Meanwhile, loudspeakers in the patient's room served up a constant stream of hortatory messages. Cameron was not a fringe figure. He was a professor at McGill University and director of Montreal's prestigious Allan Memorial Institute from 1943 to 1962. He served as president of the American Psychiatric Association in 1952–1953.

The so-called emetic therapy involved apomorphine, a synthetic morphine derivative. According to an observer, patients given apomorphine "turn green and vomit for up to an hour. This would have a sapping effect and they would finally be able to get six hours or so of needed rest." A psychiatrist at the Verdun Protestant Hospital in Montreal injected psychotic patients with turpentine, sulfur, and other toxins. Other remedies involved giving patients large doses of laxatives, pulling their teeth, and surgically removing colons, ovaries, gonads, thyroids, and other glands.

Electroconvulsive therapy was introduced in 1938 by an Italian psychiatrist, Ugo Cerletti. He found that electric current applied to the skull in sufficient strength produces a seizure resembling an epileptic fit; these seizures relieved the symptoms of some patients. Early on, patients often thrashed so violently during the seizure that they broke teeth and bones, including vertebrae. That drawback was ameliorated with straps, mouth guards, and, initially, curare, a toxin extracted from rain forest plants that in small doses produces short-term paralysis (and in larger doses death). Physicians later employed less dangerous paralytics as well as short-acting anesthetics. As a result of these improvements, electroconvulsive therapy rapidly became the "treatment of choice," as one psychiatrist put it in 1959, for severe depression and other disorders.

Even enthusiasts have acknowledged that shock treatments were abused. At the Milledgeville State Hospital in Georgia, at one time the largest asylum in the world, orderlies punished unruly inmates with what was known as the "Georgia Power Cocktail." Shock treatment declined in popularity during the 1960s, especially after being depicted as a form of torture in Ken Kesey's 1962 novel *One Flew Over the Cuckoo's Nest.* Critics of psychiatry such as Thomas Szasz, R. D. Laing, and the Scientology movement lobbied for the abolition of electroconvulsive therapy and succeeded in having it temporarily banned in certain jurisdictions. Opponents claimed that shock treatments produced permanent memory loss and other severe side effects in patients. The decline in the therapy's reputation resulted in some odd demographic facts. In 1980, not a single nonwhite American received electroconvulsive therapy at a state mental hospital. Recipients became increasingly educated, affluent, and white.

The most notorious so-called biological treatment was the lobotomy, the partial or complete destruction of the brain's frontal lobes, the seat of cognition. The technique's originator was the Portuguese neurologist Egas Moniz. In 1935, Moniz heard a lecturer describe an experiment in which a violent, uncontrollable monkey was rendered calm after its frontal lobes were "ablated," or destroyed. Moniz was soon testing the technique—in which a scalpel penetrated the brain through holes drilled into each temple—on psychiatric patients in Portugal. He reported that a majority showed improvement. Moniz received the Nobel prize in medicine and physiology in 1949. The *New York Times* noted gratefully that Moniz and other lobotomizers had "taught us to look with less awe at the brain. It is just a big organ with very difficult and complicated functions to perform and no more sacred than the liver."

On hearing of Moniz's results, the American neurologist Walter Freeman introduced lobotomies—a term he coined together with his colleague James Watts—to the United States. Freeman promulgated a variant called the transorbital lobotomy, which he performed with an ordinary icepick. The surgeon inserts the pick under the eyelid, taps the pick with a mallet until it enters the frontal lobe, and swishes it back and forth. Freeman was an indefatigable proselytizer. In one five-week period during 1951, he drove eleven thousand miles on what he called a "headhunting" expedition. Traveling in a station wagon filled with notebooks, surgical instruments, and an electroshock device, he demonstrated his transorbital lobotomy technique in hospitals throughout the United States and Canada.

The technique quickly caught on: between the late 1930s and the mid-1960s, some forty thousand lobotomies were performed in the United States alone. Lobotomies were performed not only on the mentally ill but also on incorrigible prisoners. Freeman reportedly carried out five thousand operations; he performed as many as twenty-five in a single day. He continued doing lobotomies until 1967, but by then the procedure's popularity had waned. (Freeman's last patient, upon whom he had already operated twice, died of internal bleeding.) Toward the end of Freeman's career, his reputation had become so tarnished by his zeal for brain surgery that he was stripped of his medical privileges by several institutions where he had once practiced.

Given the revulsion provoked by these approaches not only among patients and their families but also among mental health workers, psychiatrists were enormously relieved in the 1950s when they discovered medications that seemed to relieve the symptoms of some mental illnesses, and especially schizophrenia. Drugs had long been employed for the treatment of psychological ailments. In the nineteenth century, physicians prescribed morphine, chloralhydrate, and bromide salts to sedate psychotic patients.

But what is often described as the revolution in psychopharmacology began in the early 1950s with the introduction of antipsychotic medications such as reserpine and chlorpromazine (the latter marketed under the trade name Thorazine). Unlike barbiturates and other sedatives, these new drugs did not merely stupify schizophrenics but seemed to relieve their more florid symptoms, such as hallucinations. Some catatonic patients even regained their ablity to speak and move; barbiturates, if anything, deepened catatonia.

The discovery that lithium can quell the effects of mental illness dates back to the late 1940s and is credited to John Cade, the superintendent of a mental hospital in Australia. Suspecting that insanity is caused by a toxin excreted in the urine, Cade isolated various compounds from the urine of his patients and injected them into guinea pigs. After he added lithium to the compounds to make them more soluble, the guinea pigs became unusually calm. When Cade injected lithium salts into his human patients, the drug had the same calming effect. A fifty-year-old patient who had been manic for decades—Cade described him as "garrulous, euphoric, restless and unkempt"—and seemed doomed to permanent hospitalization was soon well enough to be discharged. Later studies showed that lithium abates the extreme mood swings characteristic of manic depression.

All of these medications were viewed, quite correctly, as a challenge to Freudian theory and therapy. In 1955 *Time* proclaimed that chlorpromazine and similar drugs represented the triumph of psychiatry's "red-brick pragmatists" (the term referred to the building material commonly employed to make asylums) over the "ivory-tower" psychoanalysts: "The ivory-tower critics argue that the red-brick pragmatists are not getting to the patients' 'underlying psychopathology' and so there can be no cure. These doctors want to know whether he withdrew from the world because of unconscious conflict over incestuous urges or stealing from his brother's piggy bank at the age of five. In the world of red bricks, this is like arguing about the number of angels on the head of a pin." Actually Freud had predicted just before his death that psychoanalysis might one day yield to psychopharmacology. "The future may teach us to exercise a direct influence [over the mind], by means of particular chemical substances," he wrote.

Listening to Peter Kramer

The advent of antidepressants in the 1950s posed by far the most serious challenge to psychoanalysis and other psychotherapies. Although some psychoanalysts, particularly in the United States, took on patients with schizophrenia and manic depression, Freud himself and many of his followers considered these ailments to be untreatable. The vast majority of patients treated by Freudians and other psychotherapists suffered from less severe and more common ailments, notably depression. Whereas manic depression and schizophrenia each affect roughly one out of every 100 people, depression strikes as many as 50 percent of all people at some point in their lives. At any given time, as much as 20 percent of the population may be experiencing some symptoms of depression.

The first class of drugs billed as true antidepressants were monoamine oxidase inhibitors. These compounds inhibit production of the enzyme monoamine oxidase and thus block the degradation of the neurotransmitters norepinephrine and serotonin, low levels of which are associated with depression. Unfortunately, monoamine oxidase inhibitors have a nasty side effect; when they react with tyramine, a compound found in cheese, wine, and other common foods and beverages, monoamine oxidase inhibitors can cause fatal hemorrhaging in the brain. Investigators soon identified another class of antidepressants, called tricyclics, that keep norepinephrine and serotonin levels high but do not react with tyramine.

The psychopharmacology revolution achieved its apotheosis with the introduction of selective serotonin-reuptake inhibitors, or SSRIs, in the late 1980s. After neurons emit a neurotransmitter, they usually reabsorb it through a process called reuptake. The SSRIs are thought to keep levels of serotonin high by preventing neurons from reabsorbing the neurotransmitter. SSRIs were called "selective" because—unlike their relatively "dirty" predecessors—they allegedly affect serotonin alone and not other neurotransmitters. The best-known SSRI is fluoxetine, better-known by its trade name, Prozac. Prozac is one of the great success stories of modern pharmacology—and modern marketing. Eli Lilly began selling the drug in 1988, shortly after the Food and Drug Administration (FDA) approved it. The March 26, 1990, *Newsweek* showed the green and white Prozac capsule on its cover with the headline, "A Breakthrough Drug for Depression."

When the inevitable backlash began, it was as overwrought as the positive coverage had been. In 1989, an employee of a Kentucky printing plant killed eight of his fellow employees and wounded twelve others with an assault rifle before shooting himself. After it was revealed that the man had been taking Prozac, his relatives and those of the victims sued Lilly. In 1991, the widow of the rock singer Del Shannon blamed Prozac for having driven her husband to commit suicide. *Prime Time Live, Eye on America, Geraldo Live,* and other television shows eagerly reported these cases and others in which Prozac allegedly spurred patients to commit violence against themselves or others. *Donahue* titled its program, "Prozac—Medication That Makes You Kill."

In late 1991 the FDA held hearings on the matter. Lawyers for Lilly and others contended—quite justifiably—that there was not necessarily a cause-and-effect relationship between Prozac and these relatively rare episodes of violence; a few such incidents are statistically inevitable when a drug is administered to large populations of distressed people. The FDA agreed, and controversy over the potential of Prozac for triggering violence gradually faded from public view. Most of the suits against Lilly have been settled out of court.

Psychiatrists and consumers had already demonstrated that they did not believe the scare stories, as sales of Prozac soared. By 1998, Prozac was second in annual sales only to the ulcer medication Prilosec (although sales of the impotence remedy, Viagra, seemed likely to push it to the top soon). Worldwide, more than 34 million people were taking Prozac, and sales totaled more than $2.5 billion. Incredibly, sales of Prozac continued to rise even as Lilly conceded an increasing share of its market to rival SSRIs such as Zoloft and Paxil. The market for SSRIs is growing by more than 50 percent a year, according to one estimate.

The fastest-growing segment of the U.S. market for SSRIs is children age twelve and younger—even though, according to one survey, there "has never been a double-blind, placebo-controlled study published indicating that antidepressant medications are more effective than placebo in treating child or adolescent depression." By 1997, Eli Lilly was marketing a peppermint-flavored version of Prozac for children.

The most skilled proselytizer for Prozac is Peter Kramer, an associate professor of psychiatry at Brown University who maintains a private practice in Providence, Rhode Island. Shortly after Lilly introduced Prozac in the late 1980s, Kramer began discussing the drug's implications in a column in the trade journal *Psychiatric Times*. Prozac, he said, did not merely alleviate symptoms of depression and anxiety in his patients; the drug made them much more confident and energetic. He suggested that the drug could initiate an era of "cosmetic psychopharmacology," in which patients were not only cured of their ailments but were rendered "better than well." Kramer expanded on these obervations in his 1993 book, *Listening to Prozac*, which stayed on the best-seller list of the *New York Times* for twenty-one weeks.

Kramer had become a full-fledged celebrity by the fall of 1995, when I watched him participate in a symposium on the relative merits of psychotherapy and drugs at the New School in New York City. The audience consisted primarily of women, ranging from pale-faced, black-garbed college students to Upper East Side dowagers draped in brilliant Hermes scarfs. The demographics were not surprising, given that women are twice as likely to become depressed as men.

Kramer, tall, slim, and boyish with a casual, confident manner, excelled at raising questions and toying with them without committing himself to firm answers. At one point he pondered the meaning of the intense anxiety attacks that had afflicted the psychologist William James. Were these attacks legitimate responses to James's recognition of the meaninglessness of existence, or did they stem from a chemical imbalance? Kramer could not decide. When another speaker accused Kramer of helping to foment an "epidemic" of prescriptions for Prozac, Kramer pointed out with a shrug that Freud had also helped to trigger an "epidemic" of psychoanalysis; these things happen.

Kramer insisted that he was a "great fan" of psychotherapy and had enjoyed a "long romance" with it. But his intimate relationship with psychotherapy had also helped him to see its flaws. He acknowledged that drugs might eventually "wipe out" talk therapy, for economic and social reasons. Even if drug treatments had not proved to be so effective, Kramer said, talk therapy would still be in trouble, because no one had successfully established why it works or had shown that one approach is superior to another.

One of Kramer's most revealing remarks that evening was, "I'm a great believer in ambivalence, ambiguity, fuzziness." *Listening to Prozac* was similarly noncommittal. Was Kramer pro-drugs? Well, sort of. Anti-psychotherapy? Again, sort of, but not really. Charged with being too favorable toward Prozac or hostile toward talk therapy, Kramer could point to comments in which he had expressed concern about drugs and extolled the benefits of talking cures. Like Freud, Kramer built his polemic primarily on artful descriptions of individual patients. Take Tess, a chronically depressed woman, unhappy with both her career and her love life, who had been unsuccessfully treated for years with both talk therapy and antidepressants. Two weeks after putting her on Prozac, Kramer noticed a "remarkable" transformation.

She looked different, at once more relaxed and energetic—more available—than I had seen her, as if the person hinted at in her eyes had taken over. She laughed more frequently, and the quality of her laughter was different, no longer measured but lively, even teasing.

With this new demeanor came a new social life, one that did not unfold slowly, as a result of a struggle to integrate disparate parts of the self, but seemed, rather, to appear instantly and full-blown.

"Three dates a weekend," Tess told me. "I must be wearing a sign on my forehead!"

Kramer concluded, "I had never seen a patient's social life reshaped so rapidly and dramatically. Low self-worth, competitiveness, jealousy, poor interpersonal skills, shyness, fear of intimacy—the usual causes of social awkwardness—are so deeply ingrained and so difficult to influence that ordinarily change comes slowly, if at all. But Tess blossomed all at once." Kramer cautioned that not all patients respond to Prozac in this manner: "Some are unaffected by the medicine; some merely recover from depression, as they might on any antidepressant. But a few, a substantial minority, are transformed. Like Garrison Keillor's marvelous Powdermilk biscuits, Prozac gives these patients the courage to do what needs to be done." Other psychiatrists, Kramer assured readers, had witnessed the same effect.

Kramer agonized over the philosophical implications of this new drug. Who was the "real" Tess: the unhappy, insecure woman who had suffered through a string of humiliating affairs with married men, or this reborn dynamo brimming with self-assurance? We think of ourselves as having been shaped by our experiences. What does it say about ourselves that we can be so radically transformed by a mere chemical? What does it say about the value, if any, of traditional psychotherapy? Will we lose something intrinsic to ourselves by abolishing despair and anxiety chemically, rather than through self-understanding? Is mere happiness enough?

My guess is that most readers, while mildly intrigued by the philosophical quandaries Kramer posed in Listening to Prozac, were truly gripped by his accounts of patients like Tess. I know that I was when I first read Listening to Prozac. I consider myself to be a reasonably happy person, or at least not unreasonably unhappy. But after devouring dozens of Kramer's stories about Tess and others whose lives were magically transformed by Prozac, I began to wonder whether I might benefit from the stuff. Sure, things were pretty good, but if Prozac could make them even better—boost my confidence in editorial meetings, quell the doubts that give rise to writer's block, give me more get-up-and-go as deadlines approached—why not?

When the press reported in 1994 that James Goodwin, a psychologist in Wenatchee, Oregon, had recommended Prozac for more than six hundred clients and had contended on national television that everyone could benefit from the drug, again I thought, Why not? I had a similar reaction when an article in the *New Republic*—warning that Prozac could widen the gap between haves and have-nots—proposed that Prozac be made available to the poor as well as the wealthy through a national insurance plan. Yes! Prozac for everyone!

Only later did I discover the degree to which proponents of Prozac had exaggerated its effectiveness. First, Prozac is not more effective at treating emotional disorders than older antidepressants, such as the tricyclics. In 1996 the *Journal of the American Medical Association* reported the results of a comparison of Prozac to the tricyclics desipramine and imipramine, administered to 536 depressed adults. The Prozac subjects showed slightly more improvement after one month, but the difference was not statistically significant, and even that small difference had vanished after three months. In fact, measurements at three months and six months gave a slight, albeit statistically insignificant, edge to desipramine. The investigation was, if anything, biased in favor of Prozac; it was paid for by its manufacturer, Eli Lilly.

This study and others have found some evidence for the widespread belief that Prozac produces fewer side effects than older antidepressants, but the difference is modest. A meta-analysis of forty-two separate trials by British researchers in 1994 found that the dropout rate due to side effects was 14.9 percent for Prozac and 19 percent for tricyclics. "The literature does support the clinical impression that the SSRI's are better tolerated than the [tricyclics], but the findings are not as dramatic as some clinicians might expect," a review in *Psychiatric Annals* concluded. "The overall number of patients completing treatment with these two drug classes was relatively similar."

Early reports on the SSRIs rarely mentioned what is perhaps their most significant side effect: sexual dysfunction. Eli Lilly claims that fewer than 2 percent of those enrolled in clinical trials for Prozac reported sexual disfunction, but patients rarely bring up their sexual problems unless they are specifically asked, according to Robert Segraves, a psychiatrist at Case Western University. Segraves and others have found that as many as three out of four Prozac consumers experience either a reduction in sexual desire or an impaired ability to achieve orgasm or both. In fact, Prozac has been prescribed for premature ejaculation, and at least one psychiatrist has urged that it be considered as a treatment for pedophiles and other sex offenders. (Prozac apparently has the opposite effect on bivalves. In 1998 researchers at Gettysburg College in Pennsylvania reported that mussels and clams bathed in the drug emit sperm and eggs at a higher rate.)

Peter Kramer was aware of the sexual dysfunction problem, but he relegated it to the fine print, literally, at the end of his book. He revealed in an appendix that he and other psychiatrists had observed sexual difficulties arising from Prozac "fairly often." He mused, "Here is an odd circumstance: a drug prescribed for relatively well-put-together people, often for the complaint of anhedonia [inability to feel pleasure], but well-tolerated by them even when it causes sexual disfunction." Kramer speculated that patients will tolerate "a discouraging, even embarrassing, form of impotence" caused by Prozac because the drug allows them "to experience the vibrancy of pleasure in the ordinary course of life." In Kramer's *Brave New World*, unlike Aldous Huxley's, sex is dispensable.

The Dodo Hypothesis Revisited

Perhaps the greatest myth of biopsychiatry is that antidepressants represent a tremendous advance beyond psychotherapy alone in the treatment of depression. In fact, psychotherapy and drugs produce roughly comparable outcomes, according to a 1995 report in *Professional Psychology* by David Antonuccio of the University of Nevada School of Medicine and two colleagues. The researchers found that "several meta-analyses—reported in both psychiatry and psychology journals—covering multiple studies with thousands of patients are remarkably consistent in support of the perspective that psychotherapy is at least as effective as medication in the treatment of depression." In other words, the proclamation of the Dodo in *Alice in Wonderland*—"*Everybody* has won, and all must have prizes"—applies not only to psychotherapy but also to psychopharmacology.

The conclusion has been corroborated by one of the most rigorous studies of depression ever conducted: the Treatment of Depression Collaborative Research Program, initiated by the National Institute of Mental Health (NIMH) in the late 1970s. The study involved 239 depressed patients treated for sixteen weeks at three different hospitals with one of four different methods:

- Cognitive-behavioral therapy
- Interpersonal therapy, which focuses on a patient's relationships with others
- Imipramine, a tricyclic antidepressant, plus "clinical management," a brief weekly consultation with the drug-dispensing physician that serves as a kind of placebo psychotherapy
- A placebo pill plus clinical management

The NIMH depression program offers something to please—and annoy—almost everybody. Even before the results were first published in 1989, psychologists claimed that the program had vindicated psychotherapy. In 1986 a front-page headline in the *New York Times* announced, "Psychotherapy Is as Good as Drug in Curing Depression, Study Finds." Elsewhere, advocates of cognitive-behavioral therapy asserted that it produced the best long-term results, and imipramine the worst, over the long run. Interpersonal therapists were delighted that their mode worked better than cognitive-behavioral therapy (although not quite as well as imipramine) for severely ill patients.

Proponents of antidepressants seized on the fact that a group of severely depressed patients, especially those described as "functionally impaired," seemed to respond better initially to imipramine than to either of the psychotherapies. Based on this finding, the

American Psychiatric Association recommended in 1993 that drugs be the initial treatment for depression. In fact, the major finding of the NIMH project was that all four regimes produced roughly the same results, or lack thereof. According to some measures, patients responded best to the placebo and clinical management.

The psychologist Irene Elkin of the University of Chicago, who has been more intimately involved with the NIMH project than any other investigator, reviewed its findings for the 1994 edition of *Handbook of Psychotherapy and Behavior Change.* "Although there was significant improvement from pre- to posttreatment for all treatment conditions, there were surprisingly few significant differences among the treatments at termination." She noted that for less severely depressed patients, "there was no indication of differential effectiveness among the four treatment conditions, including PLA-CM [placebo plus clinical management]." The findings suggest that "such minimal supportive therapy in the hands of an experienced therapist may be sufficient to bring about a significant reduction of depressive symptomology."

One "striking" finding, Elkin said, was the "relatively small percentage of patients who remain in treatment, fully recover, and remain completely well throughout the 18-month followup period." Only 24 percent of the patients recovered from their depression by the end of the sixteen-week treatment period and remained free of major symptoms for the next eighteen months. The percentage of patients on each regime who stayed well were as follows: cognitive-behavioral therapy, 30 percent; interpersonal therapy, 26 percent; placebo plus clinical management, 20 percent; imipramine, 19 percent.

Many mental health experts, including Peter Kramer, claim that psychotherapy and drugs work best in tandem. The author of *Listening to Prozac* once told me that he considered himself "a psychotherapist at heart" who thinks that drugs can enhance the effects of talk therapy, and vice versa. In the future, he said, "there will be something called psychotherapy that will subsume psychotherapy as it is currently practiced and psychopharmacology."

This combination approach is hardly new. In the 1920s the prominent American psychoanalyst Harry Stack Sullivan forced his patients to remain drunk for up to ten days straight before he began his talking cure.

But the view that psychotherapy plus drugs can be more effective than either drugs or psychotherapy alone has not been empirically validated. In fact, it was undermined by a large-scale survey carried out in 1995 by *Consumer Reports*, a magazine published by the the nonprofit group Consumers Union. The survey asked readers about their experiences with mental health professionals and other groups that help persons with emotional difficulties. The magazine released the results of its survey, to which four thousand readers responded, in the November 1995 issue.

The survey offered much to comfort talk therapists. Most readers said that they had been helped by psychotherapy; moreover, the longer they remained in therapy, the more they felt they had improved. Some observers worried that this finding might reflect the tendency of certain patients to become "therapy junkies." Nevertheless, the American Psychological Association immediately began using the *Consumer Reports* finding to criticize health insurers' limitations on talk therapy. Psychologists were also delighted that readers who received psychotherapy alone seemed to fare as well as those who received talk therapy plus drugs such as Prozac.

The *Consumer Reports* survey "has provided empirical validation of the effectiveness of psychotherapy," declared Martin Seligman, a psychologist at the University of Pennsylvania and former president of the American Psychology Association, in the December 1995 issue of *American Psychologist*. Seligman acknowledged that the survey had some methodological weaknesses: subscribers to *Consumer Reports* could be atypical, and those who responded to the survey even more so; moreover, there was no control group. But, he contended, these flaws were no more severe than those of more formal comparison studies.

On the other hand, the survey also lent support to the Dodo hypothesis and its corollaries: all of the therapies seemed to be equally effective, or equally ineffective. Respondents reported roughly the same degree of satisfaction whether they were treated by social workers, who require a master's degree; psychologists, who need a doctorate; or psychiatrists, who must complete medical school. Only marriage counselors scored lower than the norm. But readers reported a higher satisfaction with Alcoholics Anonymous than with any of the mental health professionals or medications.

Alcoholics Anonymous may work because it exhorts members to submit to a "higher power." The therapeutic power of religious belief was highlighted in a recent study by researchers at Duke University of eighty-seven depressed men and women aged sixty and over. Roughly half of the patients received psychotherapy, antidepressants, or a combination of both. The researchers reported in the April 1998 *American Journal of Psychiatry* that "intrinsic religiosity" was the best predictor of recovery from depression in both the treated and untreated groups. There was no evidence of significant benefits from psychotherapy, drugs, or combination therapy.

From Placebo to Panacea

Roger Greenberg and Seymour Fisher, both psychologists at the State University of New York Health Science Center in Syracuse, have been among the most persistent critics of psychopharmacology. (Greenberg and Fisher were also the authors of *Freud Scientifically Reappraised*, which I discussed in Chapter 2.) They have contended in *The Limits of Biological Treatments for Psychological Distress, From Placebo to Panacea*, and numerous articles that psychiatric medications are not as effective as advertised.

Based on their review of double-blind studies of antidepressants conducted over the past thirty years, they have concluded that the benefits of antidepressants exceed those of placebo pills by only 21 percent; that was the amount by which the improvement rate of patients taking antidepressants exceeded the improvement rate of patients taking placebos. But even this figure is suspect, according to Greenberg and Fisher, because many ostensibly double-blind studies of drugs are actually biased in favor of showing positive effects. All antidepressants usually cause side effects—such as dry mouth, sweating, constipation, and sexual dysfunction. Both patients and physicians can often determine who has received the medication, triggering an expectation of improvement that becomes self-fulfilling.

Other factors may skew results. For example, during the course of a trial, patients often drop out because of unpleasant side effects, an unwillingness to conform to the trial's protocol, or other problems. Moreover, investigators seeking subjects for a study often exclude those who seem too inarticulate or disorganized or whose depression is accompanied by other physical or mental ailments. As a result, those who complete the clinical trials are unrepresentative of the general population. Most measurements of patient outcomes are based on the assessments of clinicians rather than patients; patient ratings generally show fewer benefits. When Greenberg and Fisher examined patient ratings alone in a review of twenty-two studies, they found no advantage for antidepressants beyond the placebo effect. The implication of all these findings, Greenberg and Fisher concluded, is that "the conventional claims for the superior potency of the antidepressants have been grossly exaggerated. . . . even the relatively small advantages of antidepressants over placebo reported in the literature are not dependable."

Greenberg and Fisher have also questioned the efficacy of medications for more severe mental illnesses. Early reports on lithium—touted as a "miracle drug" and "magic bullet" for manic depression—cited cure rates as high as 90 percent, but a 1990 review by the psychiatrist Frederick Goodwin and the psychologist Kay Jamison reported more modest figures: 66 percent of those given lithium had no outbreaks of mania or depression compared to 19 percent of the control group. (The fact that Jamison is afflicted with manic depression made the endorsement particularly meaningful. Ironically, Jamison's memoir, *An Unquiet Mind*, revealed her nostalgia for her mania and her struggle to accept the

side effects of lithium, such as emotional flatness, reading difficulties, and impaired physical coordination.)

According to Greenberg and Fisher, some of the trials that
Goodwin and Jamison reviewed were flawed. A number were not
double-blind; the supervising physicians knew who was receiving
lithium and who a placebo. Other investigations included only patients whose illness had previously been "stabilized" by lithium. In
other words, trials that were supposed to determine the efficacy of
lithium included patients who had already responded favorably to
the drug.

More recent investigations of lithium show modest benefits
over a placebo or even no benefits. Long-term trials reveal that patients have difficulty coping with the side effects of lithium; about
two-thirds stop taking the drug. A review in the *British Journal of
Psychiatry* in 1995 concluded, "Unfortunately, after scrutinizing the
evidence, it seems that lithium might not be the successful pro-
phylactic that was hoped for. Psychiatrists should, therefore, reap-
praise the current consensus on the long-term treatment of
manic-depressive disorder." In fact, psychiatrists are increasingly
attempting to treat manic depression with alternative medications, notably those used to treat schizophrenia and epilepsy.

"The history of the research relating to lithium follows a familiar pattern," Greenberg and Fisher remarked:

> Once again, there is a cycle of exaggerated initial results (fos-
> tered by enthusiasm and rents in the double-blind design);
> then, increasingly more conservative reports concerning the
> magnitude of the difference between active drug and placebo;
> growing disappointment among clinicians with their everyday
> results; and heightened efforts to find alternative treatments
> that will compensate for what the original magic bullet no
> longer achieves.

Like lithium, chlorpromazine and related medications for
schizophrenia have often been described as virtual cures. But according to a leading psychiatric textbook, "a reasonable estimate

is that 20 to 30 percent" of schizophrenics taking medication "are able to lead relatively normal lives. Approximately 20 to 30 percent of patients continue to experience moderate symptoms, and 40 to 60 percent remain significantly impaired for life." Moreover, chlorpromazine and other anti-psychosis drugs often cause extrapyramidal effects, which resemble the symptoms of Parkinson's disease. Patients' movements and facial expressions become stiff and rigid; they display uncontrollable, repetitive twitching and tremors. It was in part these side effects that led psychiatrists to call anti-psychosis medications *neuroleptics*, which literally means "brain-seizing."

Extrapyramidal effects usually subside if a patient stops taking the medication. But long-term ingestion of neuroleptics can cause a more severe side effect, called tardive dyskinesia, that is usually *not* reversible. I first observed this syndrome in the mid-1970s when I met a middle-aged woman—the mother of a friend—who had been hospitalized several times for "nervous breakdowns." Her lips were permanently puckered, and she appeared to be constantly chewing something. Her hands were clenched into claws, and she repeatedly rubbed her palms with her fingertips. She seemed keenly aware of and embarrassed by her behavior; perhaps that was why she spent so much time in her bedroom. Only when I learned about tardive dyskinesia more than a decade later did I realize that her symptoms were probably a side effect of her medication. According to the National Institute of Mental Health, as many as 40 percent of those who take neuroleptics develop tardive dyskinesia.

Claims that newer neuroleptics do not trigger extrapyramidal effects and tardive dyskinesia may be exaggerated. When Johnson & Johnson introduced the anti-psychotic medication risperidone in 1994, its advertisements claimed that it produced no more extrapyramidal effects than placebos. But within a year, reports of side effects, including tardive dyskinesia, started appearing in the medical literature. Nevertheless, by 1996 risperidone had become the most widely prescribed anti-psychotic in the United States. A similar pattern occurred with the drug clozapine, which causes not only extrapyramidal effects and tardive dyskinesia—albeit at a rate lower than neuroleptics such as chlorpromazine—but also seizures and a potentially fatal blood disorder known as agranulocytosis. In 1996 Eli Lilly introduced olanzapine, which supposedly is as effective as clozapine but does not cause agranulocytosis. As one authority on neuroleptics remarked, "It is too early to tell whether these represent a true step forward or merely another false dawn."

Prescribing Placebos for Depression

At the end of *From Placebo to Panacea*, Greenberg and Fisher summarized their position on psychiatric drugs. "Are we suggesting that psychoactive drugs do not work? No, that is not our message," they stated. "The complex of ingesting a substance that palpably induces 'druglike' body experiences, in the context of personally feeling the need to change or improve, and the added element of receiving authoritative reassurance that now there is a good probability of changing—all seem to offer an opportunity for a therapeutic process to be set in motion." In other words, the placebo effect may account for much, if not all, of the beneficial effects of psychopharmacology.

Of course, as psychologists, who cannot prescribe drugs, Greenberg and Fisher may have had an allegiance to psychotherapy and a bias against medication. In their book *Freud Scientifically Reappraised* and elsewhere, they did not seem to hold psychoanalysis and other psychotherapies to the same standards of proof to which they held medications. But their assertion that the placebo effect might account for much of the effectiveness of medications for emotional disorders has been supported by Walter A. Brown, a psychiatrist at Brown University and an authority on the placebo effect.

Brown acknowledged in *Scientific American* in 1998 that the term *placebo* "brings with it connotations of deception and inauthenticity. A modern myth about placebos reflects this stigma: if a condition improves with placebos, the condition is supposedly 'all in the head.'" But research has demonstrated time and again, Brown pointed out, that patients' expectations can have very real, measurable effects. He cited one study in which asthmatic patients inhaled a mist consisting only of saltwater. When the patients were told that the mist contained allergens that could exacerbate their asthma, their lungs became more constricted and their breathing more labored; when told that the mist contained anti-asthma medication, the patients breathed more easily.

In 1994, Brown made a startling proposal in *Neuropsychopharmacology:* physicians should consider prescribing placebos as the initial treatment for many cases of depression. Brown believed that antidepressants should be prescribed for severe depression, but he was convinced by the research of Fisher, Greenberg, and others that placebo pills can be just as effective as either antidepressants or psychotherapy for many depressed patients. The major advantage of placebos, Brown pointed out, is that they are less expensive—and require less training to administer—than either active medications or psychotherapy.

Physicians need not deceive patients about the nature of the placebo for it to have a beneficial effect, Brown contended. There is evidence, albeit slight, that patients respond to a placebo even when they know that it is a placebo. In a 1965 study of fourteen depressed patients, all were prescribed placebo pills for a week. Although the subjects were all told that the pills were inert, interviews established that six believed the pills actually contained medication. Incredibly, all fourteen subjects, even those who had accepted that the pills were inert, responded positively during the week of "treatment." Four of the fourteen patients told the researchers that the placebo was the most effective medicine they had ever been prescribed; five wanted to keep taking the dummy pills after the study ended.

Brown envisioned a physician recommending the placebo treatment in the following way:

Mrs. Jones, the type of depression you have has been treated in the past with either antidepressant medicine or psychotherapy, one of the talking therapies. These two treatments are still widely used and are options for you. There is a third kind of treatment, less expensive for you and less likely to cause side effects, which also helps many people with your condition. This treatment involves taking one of these pills twice a day and coming to our office every two weeks to let us know how you're doing. These pills do not contain any drug. We don't know exactly how they work; they may trigger or stimulate the body's own healing process. We do know that your chances of improving with this treatment are quite good. If after six weeks of this treatment you're not feeling better we can try one of the other treatments.

Other psychiatrists reacted with horror to Brown's proposal. Donald Klein, a psychiatrist at Columbia University, complained that the proposal "would play into the hands of those who depict psychiatrists as artful, exploitative manipulators who take advantage of the patient's gullibility." Another psychiatrist wondered, "Will colleagues rise up to defend practitioners when the first patient managed by placebo jumps out of the window and his brother, the lawyer, arrives in court?"

Like it or not, Brown replied, many people are already seeking out their own placebo treatments in the form of homeopathic remedies and other alternative medicines that have no known pharmacological effect. "A thread that runs through alternative medicine is that the body can heal itself," Brown said. "A prescription for placebo treatment affirms this belief. And, although a treatment that works in a mysterious way may be anathema to those of us seeking rational therapies, this very mystery—and magic—may be not only acceptable but appealing to many of our patients." As Jerome Frank had pointed out in *Persuasion and Healing*, some patients respond best to nonscientific therapies.

The Resurrection of Shock Therapy

A potent symbol of the limitations of psychiatric drugs is the persistence of two notorious alternative treatments. One is the lobotomy. In his 1998 book *Last Resort*, the historian of science Jack Pressman argued that the lobotomy does not deserve its sullied reputation, given the limitations of all psychiatric remedies. In fact, psychosurgery is still occasionally prescribed as a treatment for severe mental illness that does not respond to other approaches. At Massachusetts General Hospital, which is linked to the Harvard Medical School and is one of the most prestigious hospitals in the world, surgeons offer a variant of the lobotomy, called bilateral stereotactic cingulotomy, for obsessive-compulsive disorder, manic depression, panic disorders, depression, and even substance abuse. Patients must have failed to respond to other treatments and must consent to the cingulotomy. Cingulotomy severs a marble-sized bundle of nerves, called the cingulate gyrus, that links the frontal lobes, the seat of higher cognitive functions, to the limbic system, which regulates emotion.

Initially the team inserted computer-guided electrodes into the brain through holes drilled in the skull; the electrodes burned away the tissue surrounding them with pulses of electricity. The psychosurgeons monitored their progress with MRI. The Massachusetts General group has recently replaced this apparatus with beams of gamma rays that require no drilling of the skull. When several gamma-ray beams converge on a single spot inside the brain, they destroy the tissue with far less collateral damage than previous methods. In 1996, the team reported in *Neurosurgery* that one-third of a group of thirty-four patients had shown improvement after cingulotomies.

The Massachusetts General psychosurgeons have avoided publicity, for the most part successfully (although they do maintain a Web site). But in "Lobotomy's Back," published in *Discover* in 1997, the neurosurgeon Frank Vertosick expressed ambivalence about the Massachusetts General program. He was disturbed that the researchers' assessments of the procedure's effectiveness reflected only their opinions and not those of the patients. The justification for the cingulotomy, Vertosick noted, "rests on no firmer foundation" than the cruder lobotomies performed by Walter Freeman decades earlier. Vertosick also worried that the goal of psychosurgery "wasn't to control disease but to control patients." On the other hand, he added, "some would argue that our present use of psychotropic drugs is just as flawed, in that we don't make patients better—we just succeed in preventing them from bothering us."

The other controversial treatment that has managed to persist—and even to thrive—in spite of the supposed revolution in psychopharmacology is shock therapy. One researcher who has helped to rehabilitate the reputation of shock therapy is Harold Sackheim, a psychologist at Columbia University. Sackheim works at the New York Psychiatric Institute, which is also the home of the neuroscientist Eric Kandel. Just before entering Sackheim's office, I noticed a cartoon pinned to a bulletin board beside the door. The cartoon showed a large building labeled "Institute for the Study of Emotional Stress." A man in a physician's coat had been hurled through a window and was plummeting to his death. A word balloon coming from an unseen person inside the building said, "Hey, I feel better already."

Sackheim is a slim, elegantly dressed man, with a graying moustache and watchful manner. As he gave me a tutorial on electroconvulsive therapy (commonly called ECT), he seemed to be constantly gauging my reaction to his words—no doubt because he realized that most people consider ECT to be, as he kept putting it, "horrific." When he began studying ECT in the late 1970s, its reputation was at a low ebb. "From a sociological perspective," he said, "ECT had fallen to the point where it could easily have died in America."

In 1985, however, the National Institutes of Health held a conference on ECT and concluded, "Not a single controlled study has shown another form of treatment to be superior to ECT in the short-term management of severe depression." The risk from the procedure "is not different from that associated with the use of short-acting anesthetics." In 1990 the American Psychiatric Association endorsed shock therapy for treating depression, manic depression, and schizophrenia. Shock therapy has become increasingly popular over the past decade, particularly in the United States, the United Kingdom, and Scandinavia. Sackheim estimated that it is administered to roughly 100,000 patients in the United States annually and to 1 million patients worldwide.

Just as the advent of psychiatric drugs contributed to the decline of shock therapy, Sackheim remarked, so its revival has been spurred by the growing recognition of drugs' limitations. After all, anti-psychotic medications such as chlorpromazine can cause brain damage and permanent disfigurement, as manifested by tardive dyskinesia. "But you don't get a public outcry over that." While ECT "can result in really bad memory loss, widespread, going back years," Sackheim said, the number of people who suffer such damage is much smaller than opponents of ECT have indicated.

Refinements pioneered by Sackheim's group and others also allow patients to recover quickly from ECT with fewer side effects. "I've had people star on Broadway the night after receiving treatment," Sackheim said. (His patients included "people that you see on TV every night," he added.) He and his colleagues showed that the amount of electricity required to induce a seizure varies enormously—by a factor of fifty—from individual to individual. To determine the optimal dosage for each patient, Sackheim's group starts with low current and gradually increases it. The placement of the electrodes can be critical; some patients respond best when the current is applied to only one side of the brain, others to both sides. The average course of shock therapy consists of eight or nine treatments spread out over roughly three weeks.

When I asked him whether there was a theory explaining why ECT works, Sackheim grinned broadly. There are at least a hundred such theories, he replied. (Some psychoanalysts have proposed that shock therapy fulfills patients' unconscious desire for severe punishment.) Sackheim then offered what he called the current "Columbia University gospel." The paradoxical explanation is that the brain seizure triggered by shock treatment activates the brain's intrinsic anti-seizure properties. The treatment has "profound anticonvulsant properties," Sackheim elaborated. ECT has a calming effect on the brain; after treatment, neural activity decreases, as do glucose metabolism and blood flow. If spinal fluid from an animal administered ECT is injected into other animals, it raises their seizure threshold.

Moreover, the amount of electricity required to induce seizures in humans increases "in a *huge* way" from session to session. According to this view, depression is a kind of mild, long-term seizure that can be ameliorated by an intense, short-term seizure. Sackheim compared shock therapy to stepping on a car's gas pedal when an idling engine is revving too fast. "We're triggering a seizure in order to get the brain to stop a seizure." This explanation is "probably the predominant theory right now," Sackheim said. "God knows if it's true."

In the ECT Suite

Several weeks after my initial interview with Sackheim, I met him again to observe patients being treated at the Psychiatric Institute. As we descended in an elevator from his office to the clinic, Sackheim remarked, "This is probably going to be pretty anticlimactic." Not like *One Flew Over the Cuckoo's Nest?* I asked. "No," he replied genially. In the 1975 movie Randall P. McMurphy, a character played by Jack Nicholson, feigns insanity to avoid prison and is admitted to a mental hospital. After he attacks a sadistic orderly, the orderly's colleagues strap the writhing, screaming McMurphy onto a gurney, clamp the electrodes to his temples, and shock him while he is fully conscious. Although McMurphy survives this treatment with his faculties intact, he is transformed later into a slack-jawed, blank-eyed zombie by a lobotomy. Sackheim saw the movie on his first date with his future wife, when he was still a graduate student.

Passing through a door marked "ECT Suite," we entered a small,

L-shaped room cluttered with computer monitors and other electronic gear and with people: nurses, technicians, an anesthesiologist, a psychiatrist, and two nursing students from Long Island University who had come to learn about ECT. The first patient, an elderly man wearing a pale green hospital gown, stood beside a gurney talking in a low voice to the psychiatrist, Mitch Nobler, a clean-cut young man wearing a white physician's coat.

The patient—a scientist, Sackheim informed me—lay down on the gurney. A nurse placed a white blanket over him and put the mouth guard in his mouth. Sackheim approached the patient and gave him a hearty hello, and the patient nodded. An anesthesiologist slipped a needle attached to an intravenous drug delivery system into the man's arm. Within seconds his eyes closed and his muscles sagged. Nobler, the psychiatrist, picked up what resembled a bicycle handlebar with a metal disk on one end and held it against the patient's temple. There was a brief hum. The patient became rigid, his back arched slightly, he trembled. The beeping of an electronic heart monitor accelerated. The tape extruding from a brain-wave monitor showed gentle waves yielding to sharp peaks. The seizure lasted thirty-eight seconds before abruptly ending. Within minutes, the man was groaning and shifting on the gurney. Sackheim explained that the anesthetic wears off within minutes rather than up to hours, as was the case with earlier anesthetics.

A nurse wheeled the man into an adjoining room, where patients recover. I heard someone asking him questions, to which he replied in a low, unintelligible voice. Meanwhile, another nurse rolled another gurney into the room, followed by a petite woman with short brown hair, wearing a hospital gown. She lay down on the gurney. Sackheim spoke to her softly as a technician prepared her for treatment. Coming back to my side, Sackheim explained that, unlike the elderly man, who was an outpatient and would soon go home, the woman was an inpatient. She had been admitted to the institute, Sackheim whispered, after displaying "severe impulsive suicidal tendencies."

As the assistant rubbed a conductive jelly on her temples, the woman looked increasingly fearful. Sackheim, standing at my side, told me that many patients feel anxious before their first few treatments, but their fear usually subsides over time. This was the woman's second treatment. (It had been the sixth treatment for the elderly man.) After the anesthesia rendered the woman unconscious, the psychiatrist placed two electrodes on her temples. Again I heard a hum, but unlike the previous patient the woman appeared motionless. When I asked Sackheim if the treatment had started yet, he pointed to the end of the gurney. The woman's big toe, which was protruding from the sheet, vibrated so rapidly that it blurred. The seizure lasted forty-two seconds. A few minutes later, Sackheim went to her side and said loudly, "Good morning!" She didn't respond. Her eyes remained closed. Her tongue protruded slightly from her parted lips. A nurse asked the patient, "Can you tell me your last name?" The woman mumbled her first name. She rolled over on her side. The nurse wheeled her out of the room.

Sackheim took me into the recovery room. The first patient, the elderly man, was sitting up on his gurney filling out a form that tested his cognitive functions. The woman who had just been treated lay on her side in a semifetal position as a nurse quietly spoke to her. Behind a curtain was a male patient who had been treated before Sackheim and I had arrived. When a technician asked him if he knew where he was, the patient behind the curtain mumbled incoherently. There is enormous variation in the time required for patients to recover from a treatment, Sackheim said.

Later, standing outside the entrance to the Psychiatric Institute, I asked Sackheim if he would have any misgivings about prescribing ECT for a loved one suffering from depression. "Not at all," he replied immediately, shaking his head. "Not only is the probability of getting well higher than with any other treatment," he said, "but the likelihood of getting residual symptoms is less." He lit a pipe and puffed on it contemplatively. "Personally, if I got severely depressed, I'd choose ECT," he said. "I'd want my own team to treat me," he added, "because, like all medical treatments, there is a lot of variability in how it is administered."

There is also a lot of variability in the results achieved by different groups. Published reports have indicated success rates ranging from more than 85 percent to less than 40 percent. When I asked Sackheim about his results, he said that they varied according to which type of patient he was treating. A crucial predictor of success is a patient's previous experience with antidepressant medications. Sackheim has achieved an 86 percent improvement rate with patients who have never failed to respond to an antidepressant regime.

Of course, most patients turn to shock treatments only after they have failed to respond to drugs. These patients, Sackheim said, show an initial improvement rate of only 50 to 60 percent, and 87 percent of those who improve relapse into depression within a year—and usually within four months—unless they receive more shock treatments or start responding to antidepressants. In other words, fewer than eight out of one hundred typical ECT recipients became well and stayed well without further intervention—even when treated at what may be the world's most sophisticated shock therapy clinic.

In Defense of ECT, Etc.

Various critics—notably the psychiatrist Peter Breggin, author of books such as *Toxic Psychiatry* and *Talking Back to Prozac*—have portrayed psychiatric drugs and shock therapy as evils that should be abolished. That is not my position. I share Breggin's concern that drugs and, to a much lesser extent, shock therapy have been oversold. I also agree with him that the administration of psychiatric drugs to children has gotten out of hand, especially given the absence of evidence of any benefits and the lack of knowledge about long-term negative effects. But unlike Breggin I believe that biological remedies can benefit some people some of the time, just as talk therapy can (even if the placebo effect accounts for most of the benefits). Moreover, people seeking relief from their troubles are better off having choices, even if—or particularly if—none of the choices is perfect.

Like many other people, I've had a brush with depression. In my senior year of college, a woman with whom I had been involved broke off our relationship. At the time, I was already feeling a bit anxious over what to do with myself after I graduated. What started off as ordinary misery and self-criticism rapidly mutated into something more pathological. I look different in photographs taken during that period. My facial muscles sag; my eyes look unfocused. I once heard depression described as a state of "hypervigilance." That matches my experience. I was excruciatingly aware of the passage of time, second by dreary second. My melancholy seemed literally to press down on my chest, and it never gave me a moment's respite. All the usual little pleasures—food, movies, sports, books, conversation—failed to distract me from my morbid self-scrutiny. My condition felt physiological, as much so as the flu, and it seemed to require physiological remedies—in my case, copious amounts of alcohol and drugs. But it was not *entirely* physiological. It started after a romantic relationship failed, and it ended when I met and fell in love with my future wife.

I did not seek medical help during that depression, but I probably should have. I definitely would seek help if I became depressed again; I have a family who depends on me now. I would probably try psychotherapy first, and then antidepressants. If they didn't work, and if my condition worsened, I might give the shock therapy expert Harold Sackheim a call. Because I believe that hope itself can heal, I would try my best to be hopeful. I would try to keep in mind that many cases of ordinary depression, like mine, end without any medical intervention. I would try to forget that most people who receive shock therapy for depression relapse within four months.

Given the limitations of physiological treatments for mental illness, why have they become so ascendant? Perhaps the most important factor is that they are cheaper than talk therapy (although some psychologists have disputed this claim), and they are promoted relentlessly by powerful drug companies. But the move away from psychological approaches to mental illness and toward physiological ones has also been spurred by a flood of reports that genes rather than experiences are the primary shapers of the human mind.

This genetic research has fomented a "new biological materialism," Peter Kramer asserted in *Listening to Prozac*. "When we laugh, if we do, at the claims that the genes for noticing dirty dishes, asking directions, and making commitments in relationships are absent on the Y chromosome, or that the gene for channel surfing with the TV buzz box is present only there, it is because these beliefs are not distant from the ones we actually hold." Kramer pointed out that the scientific evidence for this new genetic determinism is actually quite weak:

> It is instructive to follow the course of scientific opinion regarding the heritability of such disorders as manic-depressive illness and alcoholism. At least three times in recent years, the genes for these ailments have been discovered. In each instance, the studies proved impossible to replicate, and re-examination of the original data showed it to have been both flawed and incorrectly analyzed. My impression is that the result of each of these *failures* to demonstrate that a disorder is genetic has been a paradoxical *increase* in the conviction, both of scientists and of the informed public, that the disorder is and will be shown to be heritable in a simple, direct fashion.

This is a rather egregious case of the pot calling the kettle black, given Kramer's own exaggerations of the benefits of Prozac. On the other hand, Kramer's point is dead on: the growing conviction among scientists and the lay public that genes are the key to understanding and healing the human psyche is not merited by the research done to date.

5

GENE-WHIZ SCIENCE

. . . .

Oedipus, Schmoedipus. The Fault, Dear Sigmund, May Be in Our Genes.

—HEADLINE OF *TIME* ARTICLE REPORTING THE DIS-
COVERY OF A GENE FOR NEUROSIS

In the winter of 1993, my private and professional concerns became entangled in an unsettling way. I was beginning to gather information for an article on the fast-moving field of behavioral genetics. Through studies of twins and other related individuals, behavioral geneticists have long sought to measure the relative contributions of nature and nurture to human personality; more recently, researchers have attempted to pinpoint the specific genes underlying complex traits and disorders such as schizophrenia, sexual orientation, and high intelligence.

When I started reading up on behavioral genetics, my wife Suzie, then in her mid-thirties, was pregnant with our first child. Her obstetrician recommended that she undergo tests for birth defects. We felt fortunate that science would allow us to avoid having a child with Down's syndrome or spinal bifida or other disorders that can be identified in pregnancy. When I mentioned to a friend, another science writer, that my wife was undergoing amniocentesis, he told me that a prenatal test had had a profound effect on his life. He and his wife—I'll call them Larry and Joan—are both descended from Ashkenazi Jews, who historically have had an elevated risk of developing Tay-Sachs disease. This inherited neurological condition paralyzes children and kills them, usually after enormous suffering, before the age of five. The pattern of inheritance indicated that Tay-Sachs disease is caused by a recessive gene. The gene must be inherited from both the father and mother to be expressed; otherwise it is dormant.

In the 1970s, scientists developed tests that could determine whether someone carries the Tay-Sachs gene. Jewish organizations encouraged Jewish couples, and especially those whose families had a history of Tay-Sachs, to be tested for the disorder. When Larry and Joan decided to have children, they were tested and learned that both carried the recessive gene; each of their children would have a 25 percent chance of inheriting the Tay-Sachs gene from both parents and developing the disease. When Joan became pregnant, she was tested and learned that her fetus had inherited both genes; it was aborted. The next time she became pregnant, the tests turned out negative. Larry and Joan now have two healthy children.

The Tay-Sachs test represents, in my mind, an unalloyed scientific success. This application of genetic knowledge spared my friend and his wife—and their unborn child—terrible suffering. The same is true of thousands of other families. I cannot imagine how anyone, except perhaps the most rigid opponent of abortion, could find a downside to this scientific advance. When my wife became pregnant, behavioral genetics seemed to hold out the promise of vastly greater benefits. Researchers claimed that they could find genes not only for relatively straightforward inherited diseases—such as Tay-Sachs, Huntington's disease, muscular dystrophy, and cystic fibrosis—but also for much more complex and common disorders, such as schizophrenia, manic depression, and even alcoholism. Ultimately, scientists hoped, this genetic knowledge would yield not only prenatal tests but also better treatments and even cures.

But I ended up becoming disillusioned with the field of behavioral genetics. I was especially disturbed by the gulf between the field's modest results and the hyperbolic rhetoric that they inspired. As I was watching television one morning just before my wife had her first ultrasound test, I was startled by an advertisement for the next installment of the talk show *Donahue* (now defunct). "How to tell if your child's a serial killer!" exclaimed the announcer. When I tuned in, the host, Phil Donahue, introduced a psychiatrist who claimed to be an expert on genetics. The psychiatrist warned that men who inherit two Y chromosomes from their fathers (instead of one, which is normal) are "at special risk for antisocial, violent behavior." As evidence, he cited the case of a double-Y man in Rochester, New York, who had had an ordinary childhood but nonetheless grew up to be a serial killer; he sexually abused and strangled at least eleven women and two children before being arrested. Donahue solemnly summarized the implications: "It is not hysterical or overstating it to say that just as we are moving toward the time when, quite literally, we can anticipate or check genetic predispositions toward various physical diseases, we will also be able to pinpoint mental disorders which include aggression, antisocial behavior, and the possibility of very serious criminal activity later on."

Actually, Donahue's remarks *were* hysterical and overstated. Just before his show had aired, the National Academy of Science had issued a report on violence that stated there is *no* significant correlation between the double-Y syndrome and violent behavior. The alleged correlation had originated with British studies in the 1960s that found double-Y men to be overrepresented in prisons and mental hospitals. Investigators proposed that boys with an extra Y chromosome become unusually aggressive "supermales." But follow-up studies of nonprisoners found that double-Y men, while they tend to be taller than average and to score slightly lower on IQ tests, are not especially prone to violence.

One expects sensationalism from a television talk show host, but over the past decade, some of the world's leading scientists have indulged in equally hyperbolic rhetoric. One source of this grandiosity is the Human Genome Project. Created in the late 1980s, the project is the effort to map out all of the roughly 100,000 genes that comprise a human being. It is funded by both the U.S. government and private industry. Walter Gilbert, a geneticist and Nobel laureate at Harvard University, described the project as "the ultimate answer to the commandment 'Know thyself.'" James Watson, codiscoverer of the double helix and a former director of the Human Genome Project, told a *Time* reporter, "We used to think that our fate was in our stars. Now we know, in large part, that our fate is in our genes." The biologist Daniel Koshland, while he was editor of the prestigious journal *Science*, declared that the Human Genome Project might help to solve such seemingly intractable social problems as drug abuse, homelessness, and, yes, violent crime.

Over the past decade scientists have linked specific genes to manic depression, schizophrenia, autism, alcoholism, heroin addiction, high IQ, male homosexuality, sadness, extroversion, introversion, social skills, novelty seeking, impulsivity, attention-deficit disorder, obsessive-compulsive disorder, violent aggression, anxiety, seasonal affective disorder, pathological gambling, anorexia nervosa, and virtually every other imaginable human trait or ailment. Some critics have perceived a disturbing rightward drift behind the media's fondness for discoveries of "a gene for [fill in the blank]." But the simple fact is that these findings are classic examples of what science writers sometimes call "gee-whiz" stories; the science is fairly simple, and the philosophical and social implications are titillating. Hence the announcements keep coming.

Moreover, the political base for behavioral genetics has broadened. The effort to identify the genetic basis of character traits was once associated with social Darwinism, Nazism, eugenics, and other unsavory ideologies. To some extent, it still is. But behavioral genetics now has support from advocates for the mentally ill.

They point out that Freudian theories of mental illness—which typically blamed parents, and especially mothers, for causing schizophrenia, autism, and other disorders in their children—have needlessly compounded the anguish of parents with disturbed children. Genetic explanations of mental illness eliminate this problem, and they hold out the hope of better diagnoses and treatments. Some gay activists also support the effort to find the genetic basis of homosexuality; their hope is that society will become more tolerant of homosexuality if it is shown to stem from genes rather than a conscious choice.

Those who cling to more psychologically oriented models of the mind have struggled to come to terms with the ascendancy of the genetic paradigm. In an essay published in the *New York Times* in 1996, Adam Phillips, a child psychoanalyst, sought to find a silver lining in reports of a gene for neurosis and other traits. While accepting that these reports are probably true, Phillips suggested that psychoanalysis might still help us cope with our anxieties about genetic research. Psychoanalysis "would not and could not be in the business of disproving genetics. But it can be in the business of assessing the emotional impact of the 'acts' of genetics upon any individual person." Reading this essay, I imagined Phillips asking his hapless, supine patient, "So how did that article on the neurosis gene make you feel?" Phillips might do his anxious patient more good by pointing out that so far none of the claims linking specific genes to specific, complex behavioral traits and disorders—*not one*—has been unambiguously confirmed.

The Minnesota Twins

Behavioral genetics can be traced back to twin studies carried out in the nineteenth century by Francis Galton, a British polymath and a distant cousin of Charles Darwin. After examining both identical and fraternal twins, Galton concluded that "nature prevails enormously over nurture." In a 1865 article titled "Hereditary Talent and Breeding," he urged humanity to improve its stock through a program of selective breeding. He later described this program with the term *eugenics*, which he coined from the Greek words for "good birth."

One of the earliest and most vigorous advocates of eugenics was the American geneticist Charles Davenport. Early in the twentieth century he founded the Cold Spring Harbor Laboratory—still a preeminent biological laboratory—and the Eugenics Record Office, which gathered information on thousands of American families to track the inheritance of traits. In various publications, Davenport claimed to have demonstrated the heritability not only of eye, skin, and hair color but also of criminality, "feeblemindedness," and "pauperism." In a 1919 monograph, he asserted that the ability to be a naval officer is an inherited trait, composed of subtraits for thalassophilia, or love of the sea, and hyperkineticism, or wanderlust. Noting the paucity of female naval officers, Davenport proclaimed that the trait is unique to males.

Beginning in the 1920s, the American Eugenics Society, founded by Davenport and others, sponsored "Fitter Families Contests" at state fairs around the United States. Just as cows and sheep were appraised by judges at the fairs, so were human contestants. Less charmingly, eugenicists persuaded more than twenty U.S. states to authorize sterilization of men and women in prisons and mental hospitals, and they urged the federal government to restrict the immigration of genetically "undesirable" races. Similar practices were adopted in Canada and Europe. No other nation practiced eugenics as enthusiastically as Nazi Germany. Not content merely to sterilize undesirables, the Nazi eugenicists also practiced euthanasia ("good death") on the mentally and physically disabled—and on Jews.

As revelations of Nazi atrocities spread after World War II, popular support for eugenics programs waned in the United States and elsewhere (and support for nongenetic paradigms such as psychoanalysis rose). But the United States, Canada, and several Scandinavian countries continued forcibly sterilizing women and men believed to be genetically deficient through the 1970s; in Sweden alone, sixty thousand women were sterilized between 1935 and 1976. Twin studies and other investigations into the genetic basis of human temperament and behavior also continued. Although twin studies no longer represent the state of the art in behavioral genetics—since they cannot determine which genes contribute to behavior—they still attract an inordinate amount of attention. In fact, no other research in behavioral genetics has been embraced by the media more eagerly than the identical-twin studies done at the University of Minnesota.

The studies began in 1979, when Thomas Bouchard, a psychologist at the University of Minnesota, read a newspaper article about identical male twins who had been separated at birth and had just been reunited. The parallels between the two men were striking. Each man had been named James by his adoptive parents. James Springer had married a woman named Linda, divorced her, and remarried a woman named Betty. So had James Lewis. They had named their firstborn sons James Alan and James Allen, respectively. Each had owned a dog named Toy.

Bouchard was so fascinated by the article that he contacted Springer and Lewis and asked if he could study them. They agreed, thus initiating the Minnesota Twin Project. Bouchard eventually compiled a database on more than eight thousand twin pairs, including both identical and fraternal twins raised together. But the core of their project is identical twins raised apart, who represent a natural experiment on the relative contributions of nature and nurture to character. The researchers assume that differences between identical twins are caused by the environment; similarities are attributed to genes. The genetic component of a given trait is expressed in a term called *heritability*. Heritability applies not to individuals but to populations. The fact that height is 90 percent heritable, for example, means that 90 percent of the variation in height in a given population arises from genetic variation. The other 10 percent is accounted for by diet and other environmental factors.

By 1990 the Minnesota group had examined more than fifty

pairs of identical twins separated at birth and raised in different households. The researchers found a large genetic component in virtually every trait they studied. Whereas most other researchers had estimated the heritability of intelligence at 50 percent, Bouchard and his colleagues arrived at a figure of more than 70 percent. They also found a genetic contribution to such culturally defined traits as religiosity, political orientation (conservative versus liberal), job satisfaction, leisure-time interests, and proneness to divorce. The group summarized its findings in an article in *Science* in 1990: "On multiple measures of personality and temperament, occupational and leisure-time interests, and social attitudes, monozygotic twins reared apart are about as similar as are monozygotic twins reared together." (Identical twins are called monozygotic because they stem from a single fertilized egg, or zygote. Fraternal twins are dizygotic.)

This is one of the most disturbing statements I have ever read in a peer-reviewed article. The Minnesota group was saying, essentially, that parenting, education, and other environmental factors have little effect on who we are. The larger implication was that much of the social stratification found in the United States and elsewhere reflects genetic rather than environmental differences. Nurture doesn't really matter. Genes rule. This message was spelled out in the 1997 book *Twins* by journalist Lawrence Wright (who had previously written an excellent book on the recovered-memory movement). "The science of behavioral genetics," Wright proclaimed, "largely through twin studies, has made a persuasive case that much of our identity is stamped on us from conception; to that extent our lives seem to be pre-chosen—all we have to do is live out the script that is written in our genes."

In making this claim, Wright and other journalists have focused less on the dry heritability estimates of the Minnesota group than on the remarkable parallels between identical twins raised apart. In addition to the two Jameses, there were the "giggle" sisters. Each sister laughed constantly and wore seven rings; one had named her son Richard Andrew and the other Andrew Richard. Twins Jerry Levey and Mark Newman had each become a fireman and enjoyed Budweiser. Perhaps the most celebrated case involved Oskar, who was raised as a Nazi in Czechoslovakia, and Jack, who was raised as a Jew in Trinidad. Both were reportedly wearing shirts with epaulets when they were reunited by the Minnesota group in 1979. Each had the habit of flushing the toilet before using it and of deliberately trying to startle people by sneezing in elevators and other crowded venues.

Like Freud's case histories, these tales of the separated twins serve as a powerful rhetorical device, much more so than the statistical analyses and heritability figures. But critics have argued that the significance of these coincidences has been greatly exaggerated. In one study, the psychologist Susan Farber examined 121 cases (not including those of the Minnesota group) in which twins supposedly separated at birth were reunited later for examination by scientists. She found only three cases in which the twins had experienced no contact at all before they were studied.

Other critics have contended that if one looks for similarities between two people who look the same, were born on the same day, and were raised in the same country, chances are that one will turn up coincidences even if the two people are unrelated. Moreover, the Minnesota twins might not be representative of the general population, or even of the identical-twin population. The Minnesota group relied heavily on media coverage to recruit new twins. The twins then came to Minnesota for a week of further study—and, often, further publicity. Other studies, which maintain the privacy of the twins, have generated lower heritability estimates.

Leon Kamin, a psychologist at Northeastern University in Boston, has contended that the Minnesota twins have various motives to downplay previous contacts and exaggerate their similarities. The twins might want to please the researchers, attract more attention from the media, or even make money. There was evidence to support Kamin's suspicions. Whereas some news accounts suggested that Oskar and Jack (the Nazi and the Jew) and the two giggle sisters had been reunited for the first time in Minnesota, both pairs had met previously. James Springer and James

Lewis, after being discovered by Thomas Bouchard, appeared on the *Johnny Carson Show* and in *People* magazine. Some twins acquired agents and were paid for appearances on television and elsewhere. Oskar and Jack, after being profiled in a long article in the *Washington Post*, sold their life story to a film producer in Los Angeles. (The film was never made.)

Questions have also been raised about the motives of the Minnesota researchers. The publicity generated in the 1980s by the stories about Oskar and Jack, the two Jameses, and the giggle sisters led to a surge in funding for the Minnesota team. Its most generous backer has been the Pioneer Fund, a remnant of the American eugenics movement. The organization's name refers to its original goal of promoting the propagation of descendants of America's "pioneers"—that is, white Anglo-Saxons. The Pioneer Fund has supported various groups and scientists who advocate racial segregation.

Most of the Minnesota researchers have disavowed eugenicist policies. One exception is the psychologist David Lykken, who has proposed that the government raise the IQ of its citizenry by allowing only women who meet certain criteria to bear children; "unlicensed" women who become pregnant would be forcibly implanted with the birth-control drug Norplant. Lykken complained to the reporter Lawrence Wright, "A lot of social scientists are so scandalized by my proposals that they think I must be a fascist." A caustic reviewer of *Twins* remarked, "Now, why would anyone think that?"

Hunting Genes for Madness

Studies of twins and other family members can provide strong circumstantial evidence that a trait has a genetic component, but they cannot reveal precisely which gene or genes are involved. Advances in biotechnology in the 1980s gave behavioral geneticists the ability to scan the genes of individuals for telltale variations that might account for a given trait. If a variant of a gene, or allele, is consistently inherited with a particular trait—blue eyes, for example—then geneticists assume that the allele helps to produce blue eyes or is bound to a gene that does so.

There are two basic methods for uncovering the alleles that contribute to traits. In association studies, researchers compare the genes of unrelated individuals who share a trait to the genes of others who lack the trait. In linkage studies, investigators examine related individuals—often large, extended families or inbred ethnic groups—whose members are unusually prone to a trait. In either case, the goal is to find alleles that are consistently inherited along with the trait. In 1993, researchers used the linkage method to find an allele for Huntington's disease, a neurological disorder that usually strikes in middle age and kills within ten years. The same technique has pinpointed genes for cystic fibrosis, muscular dystrophy, and other diseases.

Like Tay-Sachs disease, these are all what are known as single-gene disorders, whose pattern of inheritance can be easily traced within a family tree. If each of two parents carries a recessive copy of the gene, their offspring have a 25 percent chance of developing the disease. If one identical twin has the disease, so will the other. The inheritance pattern of most other traits and disorders—and particularly those of interest to behavioral geneticists—is much more ambiguous. Schizophrenia, which affects 1 percent of the population, is typical. If your parent or sibling is schizophrenic, your risk of developing the disorder is 5 to 10 percent. If your identical twin is schizophrenic, you have almost a fifty-fifty chance of becoming schizophrenic as well. On the other hand, a majority of schizophrenics have no first-order relatives with the disease. Manic depression, which also afflicts about 1 in 100 people, yields similar statistics. Based on these familial data, most geneticists have concluded that schizophrenia and manic depression arise from a complex interaction of many genes and environmental factors.

Nevertheless, encouraged by the successful assaults on single-gene disorders, in the 1980s investigators began searching for the genes underlying mental illness. The potential payoff of gene-based tests and therapies for mental illness would be tremendous. Schizophrenia and manic depression each affect tens of millions of people. In 1987 geneticists claimed to have linked a gene in chromosome 11 to manic depression in an Amish population. That same year another team linked a different gene to manic depression in three Israeli families. The media hailed these findings as major breakthroughs in the field of psychiatry but gave scant notice to the subsequent failures to corroborate the initial results. A more extensive analysis of the Amish families in 1989 turned up no link between chromosome 11 and manic depression. In 1993 the results from the Israeli families were also retracted after more data were gathered.

Studies of schizophrenia have followed a remarkably similar course. In 1988 a British group announced in *Nature* that it had linked DNA in chromosome 5 to schizophrenia in families in Iceland and England. In the same issue of *Nature*, however, other investigators reported finding no linkage between the same marker and schizophrenia in a Swedish family. The British group quietly retracted its claim in 1993 after an expanded study of the Icelandic and British families found no linkage.

Peter McGuffin of the University of Wales College of Medicine, who is involved in several international efforts to identify the genes underlying mental illness, remains convinced that this research will eventually pay off. "The most powerful explanation of what's going on in bipolar disorder and schizophrenia is that they're genetic. It's just irrefutable," McGuffin told me. "There is some environment knocking around, but it's only accounting for about a fifth of the variance."

Recent linkage studies have identified several new genes that might contribute to manic depression and schizophrenia, McGuffin said. One potential manic depression gene helps to construct a receptor for the neurotransmitter serotonin, which is the target of antidepressants such as Prozac. The so-called D3 gene, implicated in schizophrenia, encodes a receptor for dopamine. Some of the most effective drugs for treating schizophrenia, McGuffin pointed out, affect dopamine levels in the brain. "I don't think it's any accident that the genetic findings are pointing somewhat in the same direction as the pharmacological findings," he said.

McGuffin was careful not to exaggerate the significance of the D3 marker and other recent findings—and with good reason: not all studies have implicated these genes in manic depression and schizophrenia. And even if the linkages hold up, McGuffin acknowledged, they might not have much practical significance. He noted, for example, that one of the most promising current candidates for a "schizophrenia gene" is the so-called 5HT2a allele. McGuffin and two colleagues conducted a meta-analysis of fifteen studies, both positive and negative, of the 5HT2a allele and found a "small but significant" linkage to schizophrenia. The analysis suggests that carriers of the 5HT2a allele have a risk of developing schizophrenia 50 percent greater than normal. Since the normal risk is about 1 percent, that translates into a risk of 1.5 percent. "At the moment, what can be done with this kind of information in terms of risk prediction is rather modest," McGuffin said.

In fact, according to McGuffin's own calculations, someone who has a single schizophrenic cousin is more than twice as likely to become schizophrenic than someone who carries the 5HT2a allele. McGuffin belongs to the Nuffield Council on Bioethics, England's main group for considering ethical issues raised by biological research. In 1998 a working group that included McGuffin released a report stating that "genetic tests for the diagnosis of the common mental disorders with more complex causes will not be particularly useful in the near future. . . . It has therefore been recommended that genetic testing for susceptibility genes which offer relatively low predictive or diagnostic certainty be discouraged until there is a clear medical benefit to the patient."

Do Prions Cause Schizophrenia?

E. Fuller Torrey is a psychiatrist at Saint Elizabeth's Hospital in Washington, D.C., and an authority on schizophrenia. In *The Death of Psychiatry, Surviving Schizophrenia, Freudian Fraud,* and other books,

Torrey excoriated theories that blame mental illness on upbringing and other psychological factors. Behavioral genetics, Torrey has asserted, is finally rendering Freudian theories of personality and mental illness obsolete. Evidence "is rapidly accumulating that genetic factors play a major role in determining many personality characteristics," Torrey contended. Behavioral genetics would have triumphed over psychoanalysis even sooner, Torrey has suggested, if not for its regrettable association with eugenics and Nazism.

There are at least two problems with Torrey's position. First, although Torrey and others often lump Freud together with hardcore behaviorists and others who adhere to a tabula rasa view of the mind, Freud himself acknowledged the importance of genetic factors. "When I laid stress on the hitherto neglected importance of the part played by the accidental impressions of early youth," he complained in 1924, "I was told that psycho-analysis was denying constitutional and hereditary factors—a thing which I had never dreamt of doing."

Second, many investigators, and notably Torrey himself, have increasingly begun to question whether genes are the key to understanding and treating mental illness. Torrey has proposed that schizophrenia may result from a virus, perhaps one that infects the fetus in utero and gradually begins damaging the brain. The pattern of incidence of schizophrenia in families often resembles the pattern exhibited by other viral diseases, such as polio, according to Torrey; if genes play a role, they may merely create a susceptibility to a viral infection. With the help of a private foundation, Torrey and other researchers at Johns Hopkins University are trying to compile evidence for the viral theory.

Meanwhile, the German virologists Liv Bode and Hanns Ludwig have presented evidence that both manic depression and common depression may stem from the so-called Borna virus, which is known to cause a neurological disorder in horses, cows, and other animals. At least one group has blamed mental illness on a completely different infectious agent called the prion. First hypothesized by Stanley Prusiner of the University of California at

Berkeley, prions are bits of protein that supposedly replicate themselves without the aid of the nucleic acids, namely, DNA and RNA.

Prions have been implicated in mad cow disease and similar human disorders that attack the brain and reduce it to a sponge-like consistency. Although Prusiner won the Nobel prize in 1997 for his work on prions, many microbiologists question whether prions actually exist. That did not stop a Brazilian team from reporting in *Nature* in 1997 that it had found evidence linking prions to schizophrenia. Other nongenetic explanations for schizophrenia blame prenatal traumas, such as poor maternal nutrition or an incompatibility between the immune system of the mother and her fetus, that damage the fetus's brain. These nongenetic explanations of mental illness are even less persuasive than the genetic ones. The fact that these alternatives are being considered at all reflects the failure of the genetic paradigm to fulfill its early promise.

The Drink Link

Is alcoholism an inherited disorder? Investigations of twins and other family members have hinted at a genetic component, especially in males who start drinking heavily early in life. But research also indicates that environmental factors play a strong role. Nevertheless, in 1990 a group led by Kenneth Blum of the University of Texas at San Antonio Health Science Center announced in the *Journal of the American Medical Association* that it had found an association between alcoholism and a stretch of DNA, or marker, near the D2 gene. The D2 gene manufactures receptors for the neurotransmitter dopamine, which has been implicated in the regulation of pleasure and a host of other mental functions.

Blum quickly applied for a patent for an alcoholism test, which he suggested parents could use to determine whether their children were at risk of alcoholism. His efforts to interest venture capitalists in such a test were no doubt aided by a front-page article in the *New York Times* portraying his study, which included thirty-five alcoholics, as a potential watershed in the diagnosis and treatment of alcoholism. The *Times* article failed to mention the considerable doubts that other geneticists had about the D2-alcoholism association.

Various researchers nonetheless took the results of Blum and his coworkers seriously enough to try to replicate them, and in 1993, the *Journal of the American Medical Association* published a review article. It found "no physiologically significant association" between the D2 marker and alcoholism. One of the authors of the review told me later that Blum's D2 claim was "a dead issue." The prominent geneticist Irving Gottesman of the University of Virginia called Blum's claim "garbage." Blum and his colleagues still insist that the D2 marker causes *something*, even if they cannot agree on what that something is. Blum's group and others have implicated the D2 allele in liver disease and other medical complications arising from alcoholism, "polysubstance" abuse (including cigarette smoking), cocaine addiction, compulsive eating, attention-deficit disorder, Tourette's syndrome, and pathological gambling. Other behavioral geneticists view the D2 affair as an embarrassment to the field that would be best forgotten.

The association between the D2 gene and alcoholism has been discounted even by Dean Hamer, one of the most energetic promoters of genetic theories of human behavior. In his 1998 book *Living with Our Genes*, which he cowrote with a journalist, Hamer held out the possibility that alcoholism genes might one day be identified, but he described Blum's D2-alcoholism association as essentially disproved. Research had shown that the allele Blum identified "was in a part of the chromosome not known to have any functional significance," Hamer remarked. "In other words, they had discovered a gene that didn't do anything." What Hamer failed to mention in his book was that his own major claim to fame—a gene associated with homosexuality—rests on even flimsier evidence than the alleged alcoholism gene.

In 1993, Hamer and four colleagues from the National Cancer Institute presented evidence in *Science* that a genetic marker within the X chromosome contributes to male homosexuality. The group performed genetic tests on forty pairs of gay brothers. If chance alone prevailed, then in only half of the pairs should both brothers have inherited the genetic marker. Instead, both siblings carried the same DNA fragment from the X chromosome in thirty-three pairs, a statistically significant result. The announcement made headlines worldwide. Hamer appeared on *Nightline* and the *MacNeil/Lehrer News Hour*, and he signed a contract to co-write a book, *The Science of Desire*, published in 1994.

In 1995 Hamer and his colleagues reported that they had replicated the X-chromosome result, but the statistical significance of the new results was not nearly as dramatic. First, the pairs of brothers in the second study totaled thirty-two rather than forty. Second, the genetic marker showed up in only twenty-two pairs, or 67 percent of the total (in comparison to the 50 percent expected from chance). The previous study had yielded a hit rate of 82 percent.

Not even this weak association has been corroborated by other researchers. In 1995 George Ebers of the University of Western Ontario examined fifty-two pairs of gay brothers and found no evidence of a linkage between homosexuality and genes within the X chromosome or elsewhere. Ebers and a colleague, George Rice, also found no support for the pattern of inheritance posited by Hamer in their study of 182 families with one or more gay males. In 1998, a group led by Alan Sanders of the National Institute of Mental Health announced that its study of fifty-four pairs of gay siblings had turned up no significant linkage between homosexuality and thirty-two different markers in chromosome X, including the marker singled out by Hamer's group.

In *Living with Our Genes*, Hamer reinterpreted the Ebers and Rice data and claimed that their work was "actually a back-handed confirmation, albeit not a significant one," of his original results. Ebers and Rice contend otherwise. (The Sanders study had not yet been publicized when Hamer wrote his book.) "The evidence is compelling that there is some gene or genes [in the X chromosome] related to male sexual orientation," Hamer concluded. In his book

Hamer highlighted two other discoveries that have enhanced his reputation as a leading gene hunter: a gene for "novelty seeking" (or "thrill seeking," as many journalists described it) and a gene for anxiety, or "neurosis." Hamer and his co-workers published both of these findings in 1996. What Hamer did not mention in *Living with Our Genes* is that at least two other groups reported in 1996 that they had found no evidence for the novelty-seeking gene.

The Bell Curve *and the Flynn Effect*

There is no more contentious component of behavioral genetics than intelligence. The controversy was instigated in large part by Cyril Burt, who in his prime was England's most influential psychologist. As the chief psychologist for the London public school system, he oversaw the testing of all its students. In the 1920s, Burt began seeking out identical twins separated at birth. He tracked down fifty-three pairs in all and followed them into adulthood. In 1966 he published the stunning results of his study. Intelligence, he found, was scarcely affected by upbringing; Burt estimated the heritability of intelligence at almost 80 percent.

In the 1970s Burt's conclusions were discredited after the psychologist Leon Kamin and others exposed inconsistencies and possible fraud in his data. Since then others have tried to rehabilitate Burt, arguing that although he may have been guilty of sloppiness and poor record keeping, his basic findings were sound. Researchers at the Minnesota Twin Project claim to have derived a heritability figure for intelligence only slightly lower than Burt's. But other geneticists favor a figure of 50 percent or even less. Bernie Devlin of the University of Pittsburgh has presented evidence that the high correlation of twins' IQ scores arises in part from their shared prenatal environment. Devlin contends that when this factor is taken into account, the heritability of intelligence falls to 34 percent.

Burt's contention that intelligence is a relatively fixed trait insensitive to environmental intervention was nonetheless resurrected in 1994 in *The Bell Curve*. (The term *bell curve* refers to the shape of a graph formed by plotting the IQ scores of a large population.) Charles Murray, a political scientist, and Richard Herrnstein, a psychologist, argued that the economic stratification of American society reflects ineradicable differences in intellectual ability; more specifically, the persistently low status of blacks relative to whites stems primarily not from discrimination or other social factors but from blacks' lower intelligence. Blacks score roughly fifteen points lower, on average, than whites on IQ tests.

Murray and Herrnstein claimed to take an agnostic stance on the question of whether this difference reflects genetic factors, but their entire polemic pointed in that direction. They argued that Head Start, improved education, affirmative action, and other programs designed to raise the status of blacks can have only marginal effects at best and are generally a waste of effort. Murray and Herrnstein were merely updating arguments made previously by others, notably by the psychologist Arthur Jensen of the University of California at Berkeley, and before them by eugenicists and social Darwinists. *The Bell Curve* nonetheless caused a sensation. It was debated on television and in the press, and it appeared on the *New York Times* best-seller list, a remarkable achievement for a book dense with statistics, charts, and graphs.

Critics raised numerous objections to *The Bell Curve*. They noted that the term *intelligence* is hopelessly vague and broad. What IQ tests measure is the ability to take IQ tests; the correlation between IQ scores and success in academia, business, and other realms is much less clear-cut than *The Bell Curve* suggests. Cognitive ability actually comes in many different forms, such as verbal, mathematical, spatial, and social. Similarly, racial categories such as *black* and *white* are not genuine biological phenomena but cultural constructs. The gap between black and white IQ scores reflects the persistent effects of racial bias in the United States rather than innate differences in intellectual ability.

None of these criticisms seemed absolutely devastating—at least, not as devastating as I would have liked. Although I believe that in general science cannot and should not be stifled, the kind of science represented by *The Bell Curve*—whether its specific assertions turned out to be right or wrong—seemed to me to have no redeeming value. Quite the contrary. The claims of Murray and Herrnstein could easily become self-fulfilling by convincing black children, their parents, and their teachers that the children are innately, immutably inferior. Why, given all the world's problems and needs, would scientists even consider questions such as these? What good could come of it? This position—that theories linking race and IQ should be ignored rather than debated—was spelled out by Noam Chomsky in *Language and Problems of Knowledge:*

> Surely people differ in their biologically determined qualities. The world would be too horrible to contemplate if they did not. But discovery of a correlation between some of these qualities is of no scientific interest and of no social significance, except to racists, sexists and the like. Those who argue that there is a correlation between race and IQ *and those who deny this claim* are contributing to racism and other disorders, because what they are saying is based on the assumption that the answer to the question makes a difference; it does not, except to racists, sexists and the like. [Italics added.]

Individuals must be considered on a case-by-case basis, Chomsky seemed to be suggesting, not on the basis of some social category to which they may belong. Unfortunately, Murray and Herrnstein made this same point in *The Bell Curve* when they argued against affirmative action and other race-based preferential programs. I was therefore relieved in 1995 when I learned of a finding that undermined a key premise of *The Bell Curve:* that intelligence as measured by IQ scores is relatively insensitive to cultural influences. Called the Flynn effect, it is named after James Flynn, a political scientist at the University of Otago in New Zealand. He stumbled on this effect in the early 1980s when he was studying the history of intelligence tests in the military.

IQ scores are ordinarily calculated by comparing an individual's performance with the performance of others in the same age group. A score of 100 is average by definition. But Flynn found that the military often recalibrated its scoring methods—or introduced new tests—to take into account a strange fact: each new generation of recruits performed better on the same test than previous generations. Soldiers who were merely average when compared with their contemporaries were above average when compared with older recruits. Put crudely, each new generation seemed to be smarter than its predecessor.

Investigating further, Flynn found that scores on virtually every kind of IQ test—administered not only to soldiers but also to students and others of all ages in at least twenty different countries— had risen roughly three points per decade for as long as such tests have existed. The gains ranged from ten points per generation, or thirty years, in Sweden and Denmark to twenty points per generation in Israel and Belgium. The upward surge tended to be greatest for tests designed to minimize cultural or educational advantages by probing the ability to recognize abstract patterns or solve other nonverbal problems. One of the most respected tests is Raven's Progressive Matrices, invented by the British psychologist J. C. Raven in 1942 and administered to people of widely varying ages since then. People tested in 1992 scored twenty-seven points higher on average than people of the same age had scored in 1942.

Flynn's data undermined some supposedly well-established claims of the intelligence-testing community. For example, many investigators had come to believe that the elderly suffer a progressive and inevitable decline in intelligence, because when they take modern IQ tests, their scores are lower than the scores of modern twenty-year-olds. But if the average seventy-year-old takes a test that was used fifty years ago, he or she will usually score as well as the average twenty-year-old of that period did on the same test. Similarly, some experts have claimed that the academic success of Chinese Americans, relative to their Caucasian contemporaries, is correlated with higher intelligence; after all, tests have shown that Chinese Americans score higher on IQ tests than other racial groups do. Flynn found that the reported IQ disparity resulted in part from the administration of old IQ tests to young Chinese Americans.

In e-mail to me, Flynn spelled out the bizarre implications of his results: "Faced with these massive IQ gains, anyone who wants to interpret them as real intelligence gains has a choice. They can either assume that the average person has normal intelligence today and that 30 to 50 years ago the average person was close to retardation, or that the average person was normal then and today the average person is close to gifted. Both assumptions are, in my opinion, absurd." No one has suggested that the Flynn effect has a genetic rather than environmental or cultural cause. As Flynn explained, "Over one or two generations, only a fanatical eugenics program could have made a significant contribution to IQ gains, and if anything mating trends have been dysgenic." The question is, What could the nongenetic cause be? Every hypothesis put forward so far has flaws, and Flynn has led the way in pointing them out.

One theory is that children have become more adept at taking tests because such tests are increasingly common. But IQ tests have actually become *less* common; moreover, studies have shown that practice has little or no effect on IQ scores, particularly for the highly abstract, nonverbal tests that exhibit the strongest Flynn effect. Attempts to correlate IQ growth with time spent in the classroom have been inconclusive; moreover, Scholastic Achievement Tests and other measures of academic accomplishment remained flat or declined in the United States even as IQ scores have risen.

Some researchers attribute the rise in IQ to childrens' increased exposure to television and other media—which have also been blamed for the "dumbing down" of modern youth. At any rate, IQ scores began rising well before the advent of television in the early 1950s. The psychologist Arthur Jensen of the University of California at Berkeley, one of the first prominent researchers to propose that blacks are innately less intelligent than whites, speculated that the IQ gains are linked to improvements in nutrition. If that were true, Flynn countered, the upward creep of IQ scores should have stalled or reversed in countries struck by famine during World War I and II.

Flynn and other psychologists see his data as a rebuttal of *The Bell Curve*. The Flynn effect highlights the vital (if mysterious) contribution of nongenetic factors to intelligence, at least as it is measured by IQ tests. The Flynn effect also suggests that the fifteen-point gap between average black and white IQ scores—far from being immutable, as Murray and Herrnstein contended—may be closable. After all, IQ has risen more than fifteen points in some countries in a single generation.

Searching for Smart Genes

At the very least, one would hope that the Flynn effect would promote an attitude of humility and modesty among intelligence researchers, especially those who insist that intelligence is a largely innate, fixed trait. One researcher who often emphasizes the limits of his work—and of behavioral genetics in general—is Robert Plomin of the Institute of Psychiatry in London. Plomin is for the most part a voice of reason and moderation within behavioral genetics, who often deplores the excesses of his colleagues. When I called him in London in 1998, he was worried that the debate between nature and nurture had swung too far toward the nature side. "I find myself now having to argue more strongly for the importance of environment," he said. After all, behavioral genetics had repeatedly demonstrated the role of nongenetic factors in shaping personality.

Yet Plomin is pursuing one of the most ambitious and controversial goals of behavioral genetics: genes for high intelligence. Plomin's methodology is simple. After categorizing schoolchildren according to IQ, he looks for alleles that are more common in the high-IQ children than in those with lower IQs. The alleles are not chosen at random; most have been previously linked to neuroreceptors or other neural components. In 1993 Plomin reported finding an excess of a specific allele in high-IQ children, but the association did not hold up in follow-up studies. (In the meantime, London's *Sunday Telegraph* had already proclaimed that Plomin had proved that "geniuses are born not made.")

In May 1998, he and two colleagues presented evidence in the journal *Psychological Science* linking a gene in chromosome 6 to high IQ in a group of 217 children. In a *New York Times* article on Plomin's finding, another psychologist commented, "I confidently predict that within two months there will be genetic centers set up for profit to test parents for this gene." In his conversation with me, Plomin doubted whether his investigations would ever yield practical applications, such as tests for high IQ or intelligence-boosting genetic treatments. Only about half of the high-IQ children carried the gene, and the apparent effect of the gene is very small. "We reckon it's certainly less than two percent of the variance, and probably more like one percent." A one percent variation is the equivalent of two IQ points. What was the purpose of his research, I asked Plomin, if it did not have any practical consequences? Finding genes correlated with high intelligence, Plomin replied, could help to illuminate the contribution of heredity to intelligence and other cognitive factors. "It's the basic-science issue of hoping to find a window in which you can look at gene-to-behavior pathways."

Plomin told me that he had just seen the movie *Gattaca* on a transatlantic flight. The film depicted a not-so-distant future in which genetic engineering could not only eliminate physical and mental illness but also boost intelligence and athletic ability. The world was ruled by the genetically enhanced; those poor souls who had not benefited from genetic engineering were known as "invalids" (with the emphasis on the second syllable). Plomin was skeptical that such scenarios would actually come to pass. "I don't think genetic engineering is at all in the cards," he said. It had proved to be extremely difficult to devise treatments even for single-gene disorders such as cystic fibrosis. "Now imagine there's one hundred genes involved, interacting with each other and with the environment. I just think it's a nonstarter to talk about genetic engineering."

Plomin is right to be modest. If history is any guide, follow-up experiments—perhaps even by Plomin himself—will fail to corroborate his discovery of a high-IQ gene, just as they have failed to corroborate all the other claims about specific genes for specific behavioral traits. But if history is any guide, these failures will not stop other scientists and journalists from portraying the field of behavioral genetics in increasingly hyperbolic terms. In *Living with Our Genes*, Dean Hamer predicted that further research would soon allow parents to select positive complex traits and eliminate negative ones before conception—just as *Gattaca* predicted. Mental illness and shyness and hyperactivity will vanish; musical talent and football-playing ability and optimism will flourish. "Whether anyone thinks it's a good idea or not, we soon will have the ability to change and manipulate human behavior through genetics," Hamer declared.

The Princeton geneticist Lee Silver cranked up the rhetoric even higher in his 1997 book, *Remaking Eden*. Silver prophesied that genetic engineering might one day divide humanity into two separate species: the Genrich class, which can afford genetic engineering, and the Natural class, which cannot. The Genrich class will be supremely talented intellectually and athletically, free of physical and mental illness, and possibly even immortal. The more brilliant these superbeings become, the more they will be able to boost their intelligence still further through new technologies, in a never-ending positive-feedback cycle. "'Intelligence' does not do justice to their cognitive abilities," Silver said of these bioengineered superbeings. "'Knowledge' does not explain the depth of their understanding of both the universe and their own consciousness."

These utopian (dystopian?) predictions are ludicrous—and, coming from leading geneticists, irresponsible—given the track record of behavioral genetics thus far. To be sure, genuine progress has been made in finding genes associated with single-gene diseases such as Huntington's chorea, cystic fibrosis, Lou Gehrig's disease, and early-onset cancers. Tests are now available for identifying those who carry these genes and thus are likely or certain to come down with the associated disease. With this knowledge in hand, researchers have begun devising therapies that can eradicate harmful genes from cells or prevent them from being expressed. More than three hundred clinical trials of gene therapy involving more than three thousand patients have been carried out around the world. As of this writing, none of them has worked.

"Except for anecdotal reports of individual patients being helped, there is still no conclusive evidence that a gene-therapy protocol has been successful in the treatment of a human disease," W. French Anderson, a leading *booster* of gene therapy research, reported in *Nature* in 1998. Some leading geneticists have even questioned whether identifying the genes that cause specific diseases will inevitably lead to better treatments. Robert Weinberg of MIT, an authority on the genetics of cancer, stated in *Science* in 1997 that for "a number of genetic diseases, knowing the genes might not help the patient one whit." If that is true of single-gene diseases, it is obviously even more true of schizophrenia, high intelligence, and shyness.

The Temperament of Jerome Kagan

My wife and I now have two children, a boy and a girl. I accept—I *know*—that temperament is innate, *to some extent.* Our destiny is encoded in our genes, *to some extent.* From the start, our son had his own distinct personality, and our daughter had hers. But their personalities keep changing too; they keep surprising my wife and me. Moreover, I cannot accept that my own role in their destiny is as negligible as some geneticists have alleged (although there are times when I wish that were true).

One of the wiser investigators of human nature whom I have met is Jerome Kagan, a professor of psychology at Harvard University. In a series of painstaking studies beginning in the 1950s, he amassed overwhelming evidence that inhibition, or shyness, and lack of inhibition, or outgoingness, are to some degree innate traits. Kagan, whom I met at Harvard in the fall of 1997, is himself a bumptious extrovert and a self-described political liberal. He admitted that he still occasionally regretted his findings. When he was in graduate school at Yale in the early 1950s, he and most of his colleagues firmly believed that "everything was nurture" and the role of biology in producing differences among people was "trivial." But then his data, "like a devil, poke up and say, 'You're wrong!'"

Environmental explanations certainly had their shortcomings, Kagan said. They could not explain why some people who have abusive upbringings become happy, healthy adults, while others who have loving, generous parents are prone to depression and other disorders. Moreover, some theories popular with psychoanalysts—such as the notion that autism is caused by "cold mothers"—were "as bad as any genetic determinism," Kagan said.

But Kagan was disturbed by some of the recent claims by behavioral geneticists. He was particularly appalled by the suggestion, made by researchers at the University of Minnesota and elsewhere, that the education, financial status, and behavior of parents have virtually no impact on a child's measured intelligence. When scientists produce a finding this "crazy," Kagan said, it is their responsibility to look for alternative explanations rather than accept the finding at face value.

Every competent biologist, Kagan said, knows that nature and nurture are often woven together so tightly that they cannot be easily unraveled, even in organisms much simpler than humans. Researchers studying the fruit fly *Drosophila* had recently discovered a pair of genes that, if inherited from both mother and father, result in wingless offspring. But when researchers raised the temperature in the laboratory by ten degrees, the genes were no longer expressed; all the offspring had wings. "That's it!" declared Kagan triumphantly. "You don't need any more examples!" He then gave me another example. Scientists had found a gene that produced high blood pressure in mice—but only if they are nursed by their biological mothers; the gene is not expressed in infants nursed by unrelated females. "Come on!" Kagan exclaimed. "It's a tapestry!"

Kagan believes in the power of science to discover regularities—even laws—governing human thought and behavior. If he didn't, he said, he would shut down his laboratory and find something else to do. On the other hand, he added, behavioral genetics, psychology, and other fields that address human nature would never match the precision and power of truly hard sciences such as celestial mechanics or nuclear physics.

This theme emerged quite clearly in Kagan's 1994 book, *Galen's Prophecy*. "I do not believe that the phenomenon of a shy or bold child will ever be understood with, or predicted from, physiological knowledge alone," Kagan wrote. "Temperamental phenomena cannot be reduced to biology." This was not wishful thinking on Kagan's part; he was forced to accept this limit of science by his own research. About one in five infants show symptoms of inhibition at birth; they respond to stimulation with obvious symptoms of distress, such as crying and thrashing their limbs. About two in five are comparatively uninhibited; they remain relaxed and sanguine when stimulated.

Yet only about half of these inhibited and uninhibited children retain these chacteristics through childhood and adolescence. Some extremely shy infants become outgoing adolescents, and some happy babies become sullen, introverted teenagers. Most children cannot be easily categorized; they display a mixture of shyness and extroversion that varies with age and surroundings. Kagan emphasized that nurture could either reinforce or diminish a trait. A mother "who consistently protected her high-reactive infant from minor stresses made it more, rather than less, difficult for the child to control an initial urge to retreat from strangers and unfamiliar events. The equally accepting mothers who made mundane, age-appropriate demands for cleanliness and conformity helped their high reactive infants tame their timidity." In other words, nurture does matter.

The Other Genetic Paradigm

Behavioral genetics is not the only gene-based paradigm that has sought to displace psychoanalysis as the primary explainer of human nature. Another contender that has attracted a great deal of attention lately is evolutionary psychology. Peering at the human psyche through the lens of Darwinian theory, evolutionary psychologists see a cluster of adaptations engineered by natural selection during our ancestors' primordial past. One might think that evolutionary psychologists and behavioral geneticists would be allies, but actually the two groups have radically different scientific outlooks and goals. Behavioral geneticists are generally unconcerned with the role that natural selection played in designing human nature, which is the *primary* concern of evolutionary psychologists. More important, behavioral geneticists focus on traits that distinguish individuals from each other, whereas evolutionary psychologists are interested in those traits that all humans share.

Evolutionary psychologists often imply that behavioral genetics is trivial, obsessed with the noise rather than the signal emerging from the symphony (or cacophony) of humanity. Our commonalities, they argue, are much more important than our differences. Evolutionary psychologists take their cue from Darwin himself. At a time when many European intellectuals viewed nonwhite races as subhuman, Darwin emphasized the unity of *Homo sapiens.* After a sojourn with the inhabitants of Tierra del Fuego at the tip of South America, he recalled being "incessantly struck" with "how similar their minds were to ours; and so it was with a full-blooded negro with whom I once happened to be intimate."

But just as modern behavioral genetics remains haunted by the specter of eugenics, so evolutionary psychology is burdened with the legacy of social Darwinism, a political ideology that essentially equated might with right. Social Darwinism is usually blamed on Darwin's contemporary Herbert Spencer, who coined the term *survival of the fittest* and upheld it as a fundamental moral principle.

Spencer repeatedly denounced social programs that aided the weak and prevented the strong from rising to the top of the social heap. "To aid the bad in multiplying is, in effect, the same as maliciously providing for our descendants a multitude of enemies," he declared in his 1874 book, *Study of Sociology*.

Even the saintly Darwin occasionally conflated *is* and *ought*. He once wrote in a letter that man "must remain subject to a severe struggle. Otherwise he would sink into indolence, and the more gifted men would not be more successful in the battle of life than the less gifted." He also expressed concern that trade unions, "which many look at as the main hope for the future, . . . exclude competition. This seems to me a great evil for the future progress of mankind."

Evolutionary psychologists have sought to distance themselves—for the most part successfully—from the less savory applications of Darwinian theory. In *Why Freud Was Wrong*, the British author Richard Webster prophesied that the new-and-improved Darwinian paradigm would eventually displace psychoanalysis once and for all as a general theory of human nature. Webster conceded that so far, evolutionary accounts of human nature leave much to be desired. Although Darwin's theory "provides a solution to the problem of species and an account of the development of organic forms," Webster wrote, "the many attempts that have been made to apply it to human nature are by no means always persuasive. While incidental insights are plentiful, Darwinian theory cannot yet offer any adequate or *comprehensive* explanation of the development of human culture or the complexity of human nature." In fact, the weaknesses of evolutionary psychology are not so different from the weaknesses of psychoanalysis.

6

DARWIN TO THE RESCUE!

. . . .

*But then arises the doubt: can the mind of man, which has,
as I fully believe, been developed from a mind as low as that
possessed by the lowest animal, be trusted when it draws
such grand conclusions?*

—CHARLES DARWIN

By day selected members of the tribe engaged in ritualized
displays of linguistic skill, striving for higher status and thus—in
the case of the males, at least—increased sexual opportunities. At
night they gathered around bonfires and imbibed fermented
juices while exchanging gossip and tribal lore. A convocation of
rain forest aborigines? No, this was the annual meeting of the Human Behavior and Evolution Society (HBES), which had assembled on the seaside campus of the University of California at Santa
Barbara, a haven of evolutionary thought. Several hundred attendees were trying to fulfill Charles Darwin's prophecy (reprinted
on the cover of the meeting's program, along with a photograph of
a bare-breasted Amazonian maiden) that "in the distant future"
psychology "will be based on a new foundation"—Darwin's own
theory of evolution by natural selection.

As usual, Darwin was right—about Darwinian psychology being in the distant future, that is. But during the 1990s evolutionary theory, like a suddenly virulent virus, began racing through the social sciences. Since the HBES was founded in 1988, it has attracted a growing number of psychologists, anthropologists, economists, historians, and others seeking to understand human affairs (in all senses of the word). Publishers have released a swarm of books by scientists and journalists propounding the "new" Darwinian paradigm, including *The Moral Animal* by the journalist Robert Wright, *The Evolution of Desire* by the psychologist David Buss, *The Red Queen* by the journalist Matt Ridley, and *How the Mind Works* by the psycholinguist Steven Pinker.

If nothing else, the HBES meeting in Santa Barbara demonstrated the astonishing versatility of what the philosopher Daniel Dennett has admiringly called "Darwin's dangerous idea." Topics ranged from the evolution of Christian symbology to the resurgence of spouse swapping among middle-class Americans. There was an amusingly self-referential quality to certain lectures. Steven Pinker began his talk on the innateness of language by remarking on how curious it was that we should gather to listen to him uttering pops, hisses, grunts, and other noises. Randolph Nesse, a psychiatrist at the University of Michigan, proposed that evolutionary psychology could help us understand *how* we understand. Geoffrey Miller, a young, long-haired psychologist, talked about how male artists, musicians, and other producers of culture often increased their sexual opportunities; meanwhile, several young women sitting in the front row watched Miller raptly.

Sex was the dominant topic inside and outside the lecture halls. During an evening beach party, a cluster of beer-swigging scientists discussed whether tender, romantic love was to some extent an evolved, innate phenomenon or a modern, purely cultural invention. One way to resolve this debate, a biologist proposed, would be to observe the sexual habits of hunter-gatherer tribes. Do males display any solicitude and tenderness toward mates before, during, or after intercourse, or is it just wham-bam-thank-you-ma'am?

Someone suggested that this question be put to an anthropologist who had sojourned with hunter-gatherers in East Africa. When fetched, the anthropologist informed his audience that members of the tribe he had lived with had generally engaged in wham-bam, "utilitarian" sex. But what would happen, he was asked, if a man in the village developed a more sophisticated sexual style, one more attentive to the needs of his female partners? Wouldn't he acquire a selective advantage? The anthropologist grinned mischievously. Anthropologists who spend time with hunter-gatherers, he said, often become very popular with the ladies of the tribe.

The meeting sounded at times less like a scientific conference than a political pep rally. The easiest way to excite the audience was to excoriate those deluded souls who insist that culture (or environment or experience) rather than biology is the primary shaper of human character and society. When the anthropologist Lee Cronk of Texas A&M University derided cultural determinism as a "religion" rather than a rational stance, his audience roared with laughter.

Watching HBES participants bonding, bickering, preening, flirting, engaging in mutual rhetorical grooming, I had to concur with their basic premise: yes, we are all animals, descendants of a vast lineage of survivors sprung from primordial pond scum. Our big, wrinkled brains were fashioned not in the last split second of civilization but during the tens and even hundreds of thousands of years preceding it. We are "stone agers in the fast lane," as one speaker put it. But once that insight is granted, the question remains: Just how much can Darwinian theory tell us about our modern, culture-steeped, complicated minds? Much of what these neo-Darwinians were saying seemed to be no more than speculation or truisms dressed up in scientific jargon.

Take the work of Devendra Singh, a psychologist at the University of Texas at Austin and a leader in the thriving field of "Darwinian aesthetics", which attempts to reveal the evolutionary logic underlying our sense of beauty. For years, Singh has been circling the globe, showing men "sexy" pictures of women—including cartoon drawings and photographs of bikini-clad models—in an effort to determine whether males share certain universal and therefore innate sexual preferences. Singh's subjects included Indian laborers who had lived in all-male camps since infancy and allegedly had never seen a real woman or even an image of one.

Singh found that while male tastes in breast size, facial structure, and other features vary between and even within cultures, all men find women with a waist-to-hip ratio of 0.7 sexually alluring. (The ratio is derived by dividing the circumference of the waist by that of the hips.) Natural selection favored this preference, Singh contended in a lecture at the HBES meeting, because a waist-to-hip ratio of 0.7 correlates so well with a woman's fertility, or "reproductive potential." But Singh's finding, translated into ordinary English, seems obvious to the point of triviality. Men desire intercourse with young, healthy women—neither starving nor obese—who have not already been impregnated by another man and whose hips are wide enough to deliver a child. Do we really need "evolutionary psychology" to tell us that?

Yes, we do! according to Leda Cosmides, a psychologist, and her husband, John Tooby, an anthropologist. Cosmides and Tooby, who founded the Center for Evolutionary Psychology at the University of California at Santa Barbara in 1994, are two leading instigators of neo-Darwinian social science. Most social scientists, they pointed out to me, still believe that our concepts of beauty are culturally determined; only Darwinian aesthetics attempts to trace our sense of beauty to its biological roots. "Waist-hip ratio was a reliable cue which over evolutionary time predicted to some extent female fertility," Tooby said. "This is something that nobody had investigated cross-culturally."

According to Cosmides and Tooby, Darwinian theory can provide a much-needed framework for psychology, anthropology, and other social sciences, which are now in disarray. In the introduction to *The Adapted Mind*, a collection of essays they co-edited with Jerome Barkow of Dalhousie University, Cosmides and Tooby wrote, "After more than a century, the social sciences are still adrift, with an enormous mass of half-digested observations, a not inconsiderable body of empirical generalizations, and a contradictory stew of ungrounded, middle-level theories expressed in a babel of incommensurate lexicons."

Only evolutionary theory, they claimed, can deliver the social sciences from this chaos. Cosmides and Tooby adopted the term *evolutionary psychology* to describe their approach. One of their central tenets is that the mind consists of numerous "modules" that, like language, have an innate basis. Natural selection designed these modules to solve problems facing our hunter-gatherer ancestors, such as finding food and shelter, acquiring a mate, raising children, and coping with rivals. The mind is not an all-purpose, computer-like machine, Cosmides and Tooby have often said, but a Swiss Army knife, crammed with different tools for different tasks.

Gender is the crucial exception to the assertion of evolutionary psychologists that all humans are born with essentially the same genetic endowment. Natural selection, Cosmides and Tooby pointed out, has constructed the minds of men and women in very different ways as a result of their divergent reproductive roles. Men, because they can in principle father a virtually infinite number of children, are much more inclined toward promiscuity than women, who because they can have at most one child per year or so are choosier in selecting a mate. Because males can never be sure that they have fathered a child, their jealousy tends to be triggered by fears of a mate's sexual infidelity; women become more upset at the thought of losing a mate's emotional commitment and thus his resources. In selecting a mate, males place a premium on youth, waist-to-hip ratio, and other physical attributes that correlate with fertility. Females, in contrast, are less concerned with the physical attributes of mates than with their "resources"—their ability to provide for a family. As one sage put it at the HBES meeting, guys like pretty girls, and girls like guys with lots of money.

Married since 1979, Cosmides and Tooby struck me when I met them as an unlikely couple. A diminutive woman with long black hair who favored miniskirts and cowboy boots, Cosmides looked as though she was headed for a night of line dancing at a country and western bar. Her bearish husband preferred a more conventional intellectual look: wire-rim glasses, basic oxford shirts, and khaki pants.

But when it comes to sheer intellectual energy and competitiveness, Cosmides and Tooby are hard to tell apart. They co-organized the HBES meeting in Santa Barbara and were ubiquitous presences there. In addition to their own talks, they introduced speakers and popped up in the audience to pose questions and deliver mini-lectures. Each speaks extremely fast, and faster still in the other's presence. During a joint interview just before the HBES meeting they constantly interrupted, annotated, and contradicted each other.

When I asked about evolutionary psychology's connection to behavioral genetics, Cosmides said, "Evolutionary psychology has very little to do with behavioral genetics. There are—"

"There are a couple of things," Tooby interjected. "One is that the [Human Behavior and Evolution Society] is extremely broad, and it has people from literature and economics and a huge range of things. So there are some behavioral geneticists in the society—"

"But John, but John," Cosmides said. Out of 120 papers being presented at the conference, she pointed out, only 2 were on behavioral genetics.

"I *understand*," Tooby replied testily. "Behavioral geneticists are an extremely small part of the society, but they *do* show up."

In spite of their disagreements, Cosmides and Tooby share an evangelical faith in the promise of evolutionary psychology. That was not to say, Tooby emphasized, that being an evolutionary psychologist is the best way to advance in academia. Many intellectuals still view any discussion of the genetic basis of human nature as "somehow an immoral enterprise. And that's just sort of the cost of doing business," he said. "It makes everything more difficult. It makes it more difficult to get yourself hired. It's more difficult to get money. But on the other hand, the productivity and the sheer illumination that this point of view gives you is worth any amount of secondary costs."

Critics of evolutionary psychology, Cosmides pointed out, often ignore the crimes committed by political regimes that viewed human nature as infinitely malleable. "They don't say, 'Oh my God, I can't possibly think of the mind as an equipotential, general-purpose machine because of the 50 million people killed by Stalin and Mao and Pol Pot in the name of the philosophy that rests on that view of mind.' It's kind of hypocritical."

"We're at the dawn of a new science," Tooby said. "It's sort of like picking up these gold nuggets, all these pieces of important illumination. But we still realize there's this vast unsettled landscape out there. We dimly perceive the outlines, but it's going to be a long time before everybody wants to rush and say, 'Gee, with this one new aspect we can now answer every question and can now write policy prescriptions with absolute confidence.'"

"And not only that," Cosmides interjected, "but social policy prescriptions rest more on your value system than any scientific finding." Evolutionary psychology in no way suggests that various human traits cannot be changed, Cosmides added. "For example, I'm nearsighted, and that's partly because of a genetic predisposition. And I can see fine. I just have to put on glasses." By helping us to understand child abuse, spousal abuse, warfare, and other undesirable behaviors, Cosmides said, evolutionary psychology can help us to reduce their incidence.

"I don't know anybody who actually understands evolutionary psychology, who really *knows* it, who finds anything threatening in it," Cosmides continued. "Some of the women evolutionary psychologists I know are some of the most radical feminists I know. And in fact they find it illuminating for understanding relationships between men and women."

"The roadblocks that are put in the way of intelligent modern evolutionary approaches to mind are basically in the long run counterproductive," Tooby concurred. "They are going to increase human suffering." Tooby's implication was clear: If rejecting evolutionary psychology harms us, then embracing it just might save us.

Steven Pinker's Way with Words

Perhaps the greatest accomplishment of Cosmides and Tooby was bringing Steven Pinker into the Darwinian fold. Pinker, director of the Center for Cognitive Neuroscience at the Massachusetts Institute of Technology, brings two special strengths to his role as an advocate for Darwinian psychology. His grounding in cognitive science, which treats the mind as a bundle of information-processing devices, adds rigor, or at least the illusion thereof, to his evolutionary explanations. He is also a skilled popularizer, whose writings appeal to scientists and laypeople alike.

I first saw Pinker in the flesh at the HBES meeting in Santa Barbara, where he summarized the major themes of his 1994 bestseller, *The Language Instinct*. Language, Pinker asserted, almost certainly was an adaptation; that is, it conferred benefits on our hunter-gatherer ancestors, allowing them to share information related to toolmaking, hunting, and other learned skills. Moreover, those most facile at language would be able to charm mates, form alliances, and enjoy other advantages that would translate into more offspring.

Pinker was cool, authoritative, witty. (To demonstrate language's ambiguity, he quoted the sentence, "Tonight, Dr. Ruth discusses sex with Dick Cavett.") The audience was standing room only. It probably didn't hurt that Pinker resembles an FM-lite rock star, with a fine-boned, almost angelic face haloed by Durer-esque curls. But Pinker can be tough. That same evening, he took on none other than Richard Dawkins of Oxford University, coiner of the term *selfish gene* and an alpha male of neo-Darwinism. In an after-dinner speech, Dawkins likened the human visual cortex to a virtual reality computer, which constructs "simulations" from limited information. Pinker, whose Ph.D. thesis at Harvard had addressed vision, jumped up from his front-row seat to find fault with Dawkins's analogy: a virtual reality machine merely transforms three-dimensional information into a two-dimensional image, but the visual cortex does the reverse, a vastly more difficult task. Dawkins, one of the fiercest polemicists of modern science, meekly granted the point.

Pinker was more or less a straightforward cognitive scientist specializing in language until 1988, when he encountered the work of Cosmides and Tooby. "I was really blown away," Pinker recalled when I interviewed him in his Cambridge apartment. "I thought it was a very impressive level of thinking and explanation." *The Language Instinct* was the first major product of Pinker's conversion to Darwinism. He then decided to write a book that would address not just language but the entire human mind. Cosmides and Tooby helped arrange for him to spend a year as a visiting scholar at the University of California at Santa Barbara. There Pinker wrote *How the Mind Works*, published in 1997. The book was not terribly original; Pinker admitted in his preface that he was building largely on the insights of others. *How the Mind Works* nonetheless represented the state of the art in Darwinian rhetoric.

We can understand the mind, Pinker asserted, only by determining what it was originally designed to do. Pinker described this concept of "reverse engineering" as follows: "In rummaging through an antique store, we may find a contraption that is inscrutable until we figure out what it was designed to do. When we realize that it is an olive-pitter, we suddenly understand that the metal ring is designed to hold the olive, and the lever lowers an X-shaped blade through one end, pushing the pit out through the other end." Once we accept that our minds—like our eyes, hands, and other features of our biology—were engineered to perpetuate our foreparents' genes, Pinker reasoned, we may gain similar insights into our thoughts, emotions, and compulsions.

Pinker leavened his book with scientific fun facts—the size of testicles of different primate species is proportional to the females' promiscuity—and with pop culture references. A discussion of the ephemerality of happiness was punctuated with a quotation from tennis star Jimmy Connors: "I hate to lose more than I like to win." Emphasizing the deep-seatedness of the male capacity for aggression, Pinker revealed that even the Dalai Lama, an avowed pacifist, loves books on warfare. To illustrate the divergent sexual preferences of males and females, he quoted the owner of a dating service: "Women really look over our profile forms; guys just look at the pictures."

But at times Pinker seemed almost too rhetorically facile. Early on in his book, he warned that there is "no shortage of bad evolutionary 'explanations.' Why do men avoid asking a stranger for directions? Because our male ancestors might have been killed if they approached a stranger. What purpose does music serve? It brings the community together." These "glib and lame" explanations have allowed critics to deride evolutionary biology as "an empty exercise in after-the-fact storytelling." But when properly done, Pinker declared, evolutionary psychology can yield theories as rigorous as any in science.

When Pinker dissected vision, memory, language, and other universal human attributes, his reverse-engineering method was indeed persuasive. But some of his hypotheses seemed rather, well, glib and lame. Toward the end of *How the Mind Works*, in a chapter titled "The Meaning of Life," Pinker "explained" virtually all aspects of modern culture, including music ("an exquisite confection crafted to tickle the sensitive spots of at least six of our mental faculties"); religion (a "desperate measure that people resort to when the stakes are high and they have exhausted the usual techniques for the causation of success"); and literature ("Fictional narratives supply us with a mental catalogue of the fatal conundrums we might face someday and the outcomes of strategies we could deploy in them. What are the options if I were to suspect that my uncle killed my father, took his position, and married my mother?" In other words, *Hamlet* is really just a survival guide).

Pinker displayed his penchant for shoot-from-the-hip speculation during our interview. He had just read an article about teenage girls—generally from affluent families—who mutilate themselves with knives, razors, fingernails, or anything else sharp. This apparently perverse, maladaptive behavior, Pinker told me, could represent a "paradoxical tactic." A girl slicing up her own forearms "is basically like a terrorist who takes a hostage that the extorted party loves," he elaborated, "except the hostage is herself." The girl must be unaware of her own motivation, so that she cannot be reasoned or punished into stopping. The girl thus forces her parents to devote more time and resources to her, enhancing her chances of reproducing later. When Pinker first divulged this theory to me, I was impressed. Only later did I realize that he had merely restated what the advice columnist Ann Landers might have said: *Some kids will do anything to get attention.*

What Noam Chomsky Really Thinks

A rather large irony overshadows Steven Pinker's career—and all of evolutionary psychology. Pinker in particular and evolutionary psychologists in general are deeply indebted to the linguist Noam Chomsky. Chomsky, who like Pinker is an MIT professor, pioneered the genetic, modular approach to the mind pursued by evolutionary psychologists. Beginning in the 1950s, Chomsky argued against the tabula rasa, inductive theory of learning set forth by John Locke in the seventeenth century and promulgated by B. F. Skinner and other behaviorists. Chomsky contended that language, arguably the trait that most distinguishes us from other animals, must be a partly innate rather than entirely learned capacity.

Unlike reading and writing, Chomsky pointed out, spoken language is common to all known cultures. All languages also share common structural features, such as verbs and nouns, which Chomsky has referred to as a "deep grammar." Perhaps Chomsky's central point was the "poverty-of-stimulus" argument: all physiologically normal children learn to speak fluently even with minimal exposure to language. Evolutionary psychologists lean heavily on the poverty-of-stimulus argument (if not the term) when they contend that many of our quirks and capacities are at least partially innate.

Yet Chomsky has been sharply critical of Darwinian accounts

of language and other aspects of the mind. Evolutionary psychologists such as Tooby and Cosmides complain that Chomsky's stance—like that of two other noted critics of Darwinian social science, the Harvard biologists Stephen Jay Gould and Richard Lewontin—is motivated by his leftist politics. But when I discussed the issue with him, Chomsky insisted that his doubts about neo-Darwinism are purely scientific. He accepted that natural selection probably played some role in the evolution of language and other human attributes. But given the enormous gap between human cognitive capacities and those of other animals, he thought that science could say little about how or why those capacities evolved.

Darwin's theory essentially says that there is "a naturalistic explanation for things," Chomsky elaborated. Anyone who does not believe in "divine intervention" accepts as much. The difficulty lies in determining what the correct naturalistic explanation *is*. Natural selection is "*a* factor in determining the distribution of traits and properties within these constraints. *A* factor, not *the* factor." Darwin himself had emphasized that nonadaptive changes also occur during evolution, Chomsky said.

Biologists could make progress in reconstructing the origins of human traits similar to those found in other animals, Chomsky said. For example, Richard Dawkins and other theorists had developed plausible computer models showing how "a flat photosensitive surface could turn into an eye in a not-too-large number of generations. But that's because you know something about the physics and physiology." The same was true of the human arm. "You can find some evidence of intermediate stages. You know something about the physics and physiology. You have homologous structures in other organisms." In the case of language and other uniquely human attributes, Chomsky said, "you don't have *any* of those things."

Chomsky noted that we utter words one at a time, in a linear fashion. But we could conceivably have acquired the ability to emit one set of sounds from the mouth and another from the nose. The ability to utter two separate sequences of noises through both the mouth and nose would provide us with a "much more complex and rich communication. We wouldn't be bound by temporal linearity." If humans had developed such a capacity, Chomsky said, evolutionary psychologists would no doubt have "explained" it as a product of natural selection. Actually, Darwinian theory neither prohibits nor demands language, nor does it constrain how the language capacity should be designed. "It doesn't predict anything!" Chomsky exclaimed.

Chomsky called evolutionary psychology a "philosophy of mind with a little bit of science thrown in." If anything, evolutionary theory can explain not too little but too much. "You find that people cooperate, you say, 'Yeah, that contributes to their genes' perpetuating.' You find that they fight, you say, 'Sure, that's obvious, because it means that their genes perpetuate and not somebody else's.' In fact, just about anything you find, you can make up some story for it."

Is Altruism an Instinct?

This problem, Chomsky remarked, also afflicted the major predecessor of evolutionary psychology, sociobiology. Evolutionary psychologists often try to distance themselves from sociobiology, which was popularized in the 1970s by the evolutionary biologist Edward Wilson of Harvard. In *Sociobiology, On Human Nature,* and other works, Wilson has argued that human social behavior is constrained, albeit more loosely, by the same evolutionary principles that determine the behavior of ants and baboons.

Stephen Jay Gould, Richard Lewontin, and other critics accused Wilson and his fellow sociobiologists of trying to resurrect social Darwinism, the nineteenth-century ideology that sought to justify racism, sexism, and imperialism by elevating survival of the fittest to a moral principle. Although the attacks on Wilson were often unfair, they were effective. Even scientists sympathetic toward sociobiology avoided using the term. When a group of Darwinians founded the Human Behavior and Evolution Society in 1988, they deliberately chose not to include *sociobiology* in the society's name. HBESers adopted a preexisting quarterly, *Ethology and Sociobiology*, originally founded in 1981, as its flagship journal, but they changed the name in 1996 to *Evolution and Human Behavior*.

Ironically, that same year the HBES invited Edward Wilson to give the keynote address at its annual meeting, and Wilson took the opportunity to scold the HBES leaders for their "failure of nerve." By rejecting the term *sociobiology*, Wilson complained, they were implicitly acceding to the charge that it was a "racist, determinist" ideology. Some HBES leaders have admitted that they hoped to avoid the tainted political connotations of sociobiology. Sociobiology "raised too many hackles and got us into too much trouble," an editor of *Evolution and Human Behavior* told a reporter for *Science*.

Some evolutionary psychologists have insisted that their approach to the mind differs in significant ways from sociobiology. One distinction, according to Cosmides and Tooby, is that evolutionary psychologists assume that the mind is adapted not to modern life but to the conditions in which it evolved, "especially hunter-gatherer environments." Some sociobiologists made this assumption, but others assumed "that behavior is everywhere adaptive, even in modern environments," Cosmides and Tooby argued. These distinctions anger veteran sociobiologists such as Richard Alexander of the University of Michigan. He and other sociobiologists have never held that humans strive to maximize their reproductive opportunities in all environments, Alexander asserted. "I don't have anything against them calling themselves evolutionary psychologists," Alexander said of Cosmides and Tooby. "What ticks me off is when you have to create a straw man that you say you're superseding."

The fact is that evolutionary psychology rests on the same theoretical foundation as sociobiology. One notable achievement of sociobiology was its explanation of altruism, which is defined as behavior that benefits others at some cost to the altruist's fitness, or reproductive potential. How could natural selection, which ruthlessly rewards the most selfish of genes, favor such behavior?

Darwin himself speculated that natural selection might allow the emergence of altruistic behavior if it benefited the entire group to which the individual belongs. But so-called group selection fell out of fashion in the 1960s when the evolutionary theorist George Williams and others showed that mathematically it made no sense; genes for pure altruism would almost invariably vanish under the pressure of natural selection.

Shortly thereafter the British biologist William Hamilton proposed the theory of kin selection, which posits that natural selection will favor the emergence of altruistic behavior if that behavior enhances the reproductive prospects of the organism's kin. Kin selection revealed the evolutionary logic underlying the behavior of such highly social creatures as ants, termites, and mole rats. It could also explain the extraordinary risks and sacrifices that human mothers or fathers will make to ensure the survival of their children or other close relatives. (Asked if he would lay down his life for his brother, the British biologist J. B. S. Haldane quipped, "No, but for two brothers or eight cousins.")

What about good Samaritans, who risk their lives for total strangers? A solution to this conundrum was proposed in the 1970s by Robert Trivers of Rutgers University, another evolutionary theorist to whom evolutionary psychologists are deeply indebted. Trivers contended that natural selection might have favored the emergence of altruistic sentiments and behavior toward non-kin if such behavior resulted in a net, aggregate benefit for altruists. The good Samaritan risks his life to save a stranger from thieves, and then the stranger rewards the Samaritan with gifts of livestock and gold. The Samaritan may have acted out of genuine compassion and generosity, but he is actually carrying out a selfish, tit-for-tat strategy designed to propagate his genes. Trivers called this mechanism *reciprocal altruism*.

Skeptics have wondered why, if altruism toward non-kin is encoded in our genes, it is such a fragile feature of human psychology and history? And why must so many cultural institutions devote so much energy to indoctrinating citizens about the value of compassion and generosity? "Why so many churches, priests, judges, rabbis, probation officers and good citizenship certificates?" H. Allen Orr, a biologist at the University of Rochester, asked in a recent essay on altruism. Orr suggested that altruistic behavior, or "virtue," may be "one of these abiological, unnaturally-selected characteristics arrived at not by genes, but by hard-won experience of what does and doesn't work in human society."

Trivers's reciprocal altruism hypothesis has nonetheless become a keystone of evolutionary psychology. Leda Cosmides has argued that natural selection might have bequeathed us an innate instinct for sniffing out "cheaters"—those who might deceive us in reciprocal, tit-for-tat exchanges. In a series of experiments, Cosmides showed that people are much more adept at solving problems if those problems are posed in the context of social exchanges—and particularly ones that lead people to suspect they are being deceived—rather than as purely logical exercises. Invoking Chomsky's poverty-of-stimulus argument, Cosmides concluded that our sensitivity to potential cheaters is too exquisite to have been acquired through experience alone.

Doubts about the cheater-detection hypothesis were raised even at the HBES meeting in Santa Barbara. James Fetzer, a philosopher at the University of Minnesota, noted that we might be more skilled at determining when someone is deceiving us than at solving problems in logic simply because we encounter more situations of the former type. This and many other human abilities, Fetzer contended, might stem from a general-purpose intelligence or learning program—which relies heavily on heuristics, or trial and error—rather than innate, dedicated "modules."

Much the same argument was made at the HBES meeting by Steven Mithen, a British anthropologist. He complained that evolutionary psychologists often suggest that our ancestors evolved under more or less fixed circumstances, which is sometimes called the environment of evolutionary adaptation. Actually, Mithen pointed out, the environment in which our ancestors evolved was extremely variable and volatile. Given this fact, natural selection might have favored the emergence of a flexible problem-solving ability in addition to modules dedicated to specific tasks. Indeed, science itself—and many other aspects of human civilization—testifies to the human capacity for solving problems of all types.

Even sexual behavior, arguably the most instinctual aspect of human nature, might often be more rational than evolutionary psychologists contend. Evolutionary psychologists such as David Buss of the University of Michigan assert that females are *instinctively* more coy than males and *instinctively* place more emphasis than males on mates' resources rather than purely physical attributes. But as Buss himself acknowledges, female sexual behavior is enormously variable; some females are highly promiscuous, and not all prefer affluent men over physically attractive ones.

Let's assume for a moment that the evolutionary psychologists are right about female sexual preferences. Are these preferences really *instinctual*? It seems just as plausible—if not more so—that these preferences derive from rational, conscious deliberation. By puberty most females recognize that even if they employ contraception, they are at risk of becoming pregnant during a sexual encounter; it is thus quite rational for females to be more wary of casual sex than males are. Similarly, the female preference for males with resources might simply reflect females' rational recognition of their relatively precarious economic status and prospects.

The Evil-Father Syndrome

Evolutionary psychology is weakest when it attempts to explain unusual human behavior, such as the murder of children by their parents. To Darwinians, who view procreation as the sine qua non of life, these are the most perverse of all acts. The problem was taken up in the 1980s by Margo Wilson and her husband, Martin Daly of McMaster University in Canada, among the most respected of all evolutionary psychologists. After analyzing murder records from the United States and Canada, Wilson and Daly determined that children were roughly sixty times more likely to be killed by a stepparent—and usually a stepfather—than by a natural parent. They pointed out that this type of non-kin infanticide is common in nature; males of many species, from mice to monkeys, kill offspring that their mates conceived with another male.

The selfish-gene perspective was upheld after all. Or was it? Even Wilson and Daly have warned that their results should be interpreted with caution—and with good reason. Clearly one cannot say that men have an innate propensity to kill their mate's children if the children were fathered by other men, because the vast majority of stepfathers never abuse or kill their children. Moreover, fathers who adopt children are even less likely than biological fathers to kill or abuse their children. Of course, men who adopt children are atypical, because they are screened for emotional and financial stability—but that is exactly the point. Men who abuse stepchildren are obviously atypical too. They may have assumed responsibility for a spouse's children reluctantly. They may be subject to unusual financial and emotional stresses. These are the factors that lead certain men to kill or harm a mate's children—not some instinctual urge that they share with mice or monkeys.

Wilson and Daly's research is nonetheless often cited as a model of Darwinian social science, since it addresses an important issue and rests on a large empirical foundation. When the *New York Times* in 1997 asked leading intellectuals to name the last book they had read twice, Steven Pinker singled out *Homicide*, a book in which Wilson and Daly presented an evolutionary view of human violence. Ironically, Pinker later wrote an article for the *New York Times Magazine* that inadvertently undermined the Daly and Wilson research—and, indeed, the entire enterprise of Darwinian psychology.

Pinker's article addressed a spate of incidents in which biological mothers had killed their newborn children. (In one case, a girl at a high school dance gave birth in a bathroom stall, killed the infant, and then returned to the dance floor.) Although maternal infanticide seems at first glance to be the ultimate violation of

Darwinian precepts, Pinker said, it might result from natural selection. He noted that in certain stressful circumstances, our maternal ancestors would have been well advised to kill a newborn baby rather than devoting scarce resources to it, resources needed to sustain the mother and her older offspring. This innate psychological module might be switched on in modern mothers by severe stress.

A few weeks after Pinker's article was published, the *New York Times Magazine* printed a letter from Claude Fischer, a sociologist at the University of California at Berkeley. Pinker's article, Fischer complained, "illustrates how silly evolutionary explanations of human behavior have become. When mothers protect their newborns (which almost all do), it's because that behavior is evolutionarily adaptive. And now, when a few mothers kill their newborns, that's evolutionarily adaptive too. Any behavior and its opposite is 'explained' by evolutionary selection.... Thus, nothing is explained."

Avoiding Behavioral Genetics

Evolutionary psychology would be enormously valuable if it could predict in advance which particular parent or stepparent will abuse or kill a child, but evolutionary psychology offers no special insights into this issue. In response to the question, "Why did this woman kill her child?" evolutionary psychology replies, "All women have an innate capacity to kill their infants under certain stressful circumstances." Even if this supposition is true, it is moot, because the vast majority of women do *not* kill their children under stressful circumstances.

One possible explanation of infanticide is that certain mothers have a genetic predisposition to severe postpartum depression. That is what a behavioral geneticist might suggest. But as I mentioned at the end of the previous chapter, evolutionary psychologists often (although certainly not always) distance themselves from these sorts of explanations. "Heritable behavioral variation is in general neither predicted by nor supportive of adaptationist theories," Martin Daly, coauthor of the evil-stepparent hypothesis, once wrote. Similarly, Cosmides and Tooby have speculated that genetic variation evolved as a defense against parasites. These variations, they contended, may have minor psychological or behavioral consequences; most of the psychological and behavioral differences that distinguish individuals are due to environmental factors.

Evolutionary psychologists have two motives for making this assumption. First, if most commonalities can be ascribed to genes and most disparities to environment, the task of constructing models of human nature is enormously simplified. Second, this assumption allows evolutionary psychologists to adopt the classic liberal principle that environmental rather than innate differences may account for many of the observed differences between individuals and, more important, racial groups. Evolutionary psychologists can thus avoid the criticism that has tarred works such as The Bell Curve.

But evolutionary psychologists who disavow behavioral genetics are being both disingenuous and inconsistent. Without genetic variation among individuals, natural selection would lack the material necessary to work its magic; evolution could not occur. Furthermore, if genes can account for our commonalities, as evolutionary psychologists argue, surely genes can also account for our differences. Although the link between specific genetic differences and specific behavioral differences is far from established (as I tried to demonstrate in the previous chapter), it cannot be ruled out.

Behavioral genetics, for all its faults, at least has some hope of providing clinically useful information about pathological behavior and cognition. Not so evolutionary psychology. In How the Mind Works, Steven Pinker ignored mental illness (with the exception of autism), perhaps wisely. In their 1994 book Why We Get Sick: The New Science of Darwinian Medicine, the evolutionary biologist George Williams and the psychiatrist Randolph Nesse, two eminent members of the neo-Darwinian movement, speculated that schizophrenia, depression, panic attacks, and other disorders have persisted because they conferred some benefits on our ancestors. Schizophrenia, for example, might "increase creativity or sharpen a person's intuitions about what others are thinking."

Even less persuasive examples were presented in the 1996 book *Evolutionary Psychiatry: A New Beginning*, by Anthony Stevens, a Jungian analyst, and John Price, a psychiatrist. Women are more prone than men to agoraphobia (fear of open spaces), Stevens and Price surmised, because our ancestral mothers stayed at home with the kids while the men were out hunting. Obsessive-compulsive disorder might be a relic of our ancestors' habit of constantly checking the status of fences and other defenses against predators and rivals. Homosexuality might have persisted because homosexuals helped relatives raise their children, thus promoting the family's "inclusive fitness." To support this last hypothesis, Stevens and Price claimed that wealthy modern homosexuals "sometimes shock their friends by leaving the bulk of their estate . . . to some nephew or niece whom they may not have seen in years."

"Alas, poor Darwin!" the British biologist Steven Rose wailed in his review of *Evolutionary Psychiatry* in *Nature*. "More idle speculation and dogmatic assertion have been published in your name in the past two decades than in the full preceding century, and still the torrent continues." Rose added, "At a period when both evolutionary theory and psychiatry are confronted by entrenched cultural opponents, they deserve better than this from their friends."

Darwinian Culturalists

Ironically, in their effort to avoid becoming tarred as genetic determinists, some prominent neo-Darwinians have become virtually indistinguishable from their putative archrivals, the cultural determinists, who emphasize cultural influences on human nature. One striking example is the anthropologist Napoleon Chagnon, who like Cosmides and Tooby teaches at the University of California at Santa Barbara. Chagnon, an HBES cofounder and former president, is a swashbuckling man who enjoys recounting his many feats of daring among the Yanomamo, an Amazonian tribe that he began studying in the 1960s.

In this polygynous society, one of the few still clinging to its primordial way of life, men in one village often raid other villages, killing men they encounter there and kidnapping females. Men in the same village also fight each other, again often in disputes over women. In one form of duel, two men take turns striking each other on the head with enormous clubs until one is knocked out or otherwise disabled. Men display the lumps and scars resulting from these duels like badges of honor. The most dramatic finding to emerge from Chagnon's years in the jungle was that males who kill the most also have the most offspring. Conversely, those who shrink from violent encounters—whom I once heard Chagnon call "wimps"—have relatively few children or none.

Chagnon's discovery has chilling implications. If there are genes that predispose certain men to violence, natural selection might favor those genes in Yanomamo-type cultures, in which violence correlates strongly with reproductive success. But Chagnon, at least when I spoke to him, resisted this interpretation. He emphasized that he did not believe, as some critics and journalists have charged, that Yanomamo males or even all males have a "warfare gene." The Yanomamo men, Chagnon suggested, engage in aggressive behavior not because they are innately violent but because violent behavior is esteemed by their culture. The leaders of Yanomamo villages employ violence in a controlled manner; males who cannot control their aggression generally do not live long enough to father children.

If Yanomamo males were raised in a society that esteemed not fighting ability but farming skills, Chagnon said, they would quickly conform to that system. I suggested to Chagnon that his explanation sounded like something that the evolutionary biologist Stephen Jay Gould of Harvard might say; Gould, after all, has also emphasized the malleability of human nature (and is considered by most evolutionary psychologists as an archenemy of their enterprise). I meant to goad Chagnon with the comparison, but to my surprise, he did not object. "Steve Gould and I probably agree on a lot of things," Chagnon replied serenely.

Evolutionary psychologists (or anthropologists, in the case of Chagnon) are wise to acknowledge the tendency of people to conform to their culture. How else could they explain the enormous variety of social behavior between and even within cultures? Japan was an extremely aggressive, war-prone society during the first half of the twentieth century, but since World War II it has been a nation of pacifists. Unfortunately, conformity also poses a problem for evolutionary psychology. To demonstrate that a trait is innate, evolutionary psychologists try to show that it occurs in all cultures. In this way, for example, evolutionary psychologists have sought to establish through cross-cultural studies that males are inherently more inclined toward promiscuity than females.

But given the interconnectedness of almost all cultures, even so-called primitive ones, in the late twentieth century, some of the universal, "instinctual" attitudes and actions documented by Darwinian researchers might actually result from conformity. That is what the cultural determinists have said all along. Evolutionary psychologists such as Cosmides and Tooby suggest that genes underlie our commonalities and the environment our differences. But the reverse may also be true, at least to some extent: culture may account for many of our commonalities and genetic variation for many of our differences.

The Birth-Order Gambit

The aversion of evolutionary psychologists to behavioral genetics may explain why so many of them have embraced the birth-order hypothesis of Frank Sulloway, a historian at the Massachusetts Institute of Technology. Drawing on Darwinian theory, Sulloway has suggested how a purely experiential phenomenon—the order in which a child is born relative to siblings—can give rise to profound, permanent differences in personality. According to

Sulloway, firstborn children are much more likely than their younger siblings to be conservative, support the status quo, and reject new scientific or political ideas. Later-born children tend to be more adventurous, radical, open-minded, willing to take risks. (Sulloway, naturally, is the youngest of three brothers.)

A theory advanced by Robert Trivers, Sulloway contended, accounts for these findings. Trivers pointed out that with the exception of identical twins, siblings share only 50 percent of each other's genes. While they should be more generous toward each other than toward non-kin, siblings should also compete for parental affection and other "resources." The longer that children survive beyond the perils of infancy, the more likely they are to reproduce and propagate their parents' genes (other factors being equal). Hence parents are likely to invest more affection and resources in older children.

This fact, Sulloway contended, necessitates different strategies for children born at different times. Firstborns should maintain a close relationship with their parents and not challenge their authority. Later-borns, with less to lose, have more incentive to embrace change and disorder for their own sakes. "From a Darwinian point of view, it is just impossible that birth-order effects don't exist," Sulloway declared in a speech at the HBES meeting.

Sulloway said that these conclusions are supported by scores of studies. He also claimed to have compiled "overwhelming" evidence that most of the great revolutions in modern history—scientific and political—have been led and supported by later-borns and opposed by "stubborn" firstborns. Darwin, for example, was the seventh of eight children, and those who supported his theory also tended to be later-borns, according to Sulloway's analysis. Other huge historical shifts instigated primarily by later-borns were the Reformation and the Copernican revolution.

Sulloway had explanations for all the exceptions to his rule. Martin Luther, a firstborn, came from the lower ranks of the church and thus was able to overcome his conservatism. Newton was a firstborn, but his father died before he was born; he also hated his stepfather, thus negating the close parental bond typical of most firstborns. Freud was a firstborn, but as a Jew in an anti-Semitic society he had many of the characteristics of a later-born. The French Revolution was largely carried out by firstborns, such as Robespierre. But the French Revolution was noteworthy for its bloodiness and brutality, Sulloway explained, and these are first-born characteristics.

Born to Rebel, in which Sulloway spelled out his theory, was widely acclaimed after its publication in 1996. Sulloway's former teacher Edward Wilson called it "one of the most authoritative and important treatises in the history of the social sciences." The anthropologist Sarah Blaffer Hrdy predicted that Sulloway's work "will have the same kind of impact as Freud's and Darwin's." One expects book jacket blurbs from colleagues to be excessive, but coverage in the *New Yorker, Newsweek,* and other publications was almost equally hyperbolic. Even Stephen Jay Gould, who else-where has excoriated evolutionary psychology, praised Sulloway's work—which, after all, gives as much emphasis to nurture (albeit in Darwinian guise) as to nature in shaping personality.

But just because Sulloway's thesis appeals to scientists of all po-litical stripes does not mean that it is right. Sulloway's conclusions contradicted those contained in the 1983 book *Birth Order: Its Influ-ence on Personality.* The authors, the Swiss psychiatrists Cecile Ernst and Jules Angst, sifted through hundreds of previous studies at-tempting to link birth order to personality traits, and then con-ducted their own survey of 7,582 college-age residents of Zurich. They concluded not only that birth-order effects do not exist but also that continued efforts to find such effects represent "*a sheer waste of time and money.*"

Sulloway was well into his birth-order project when he came upon the book by Ernst and Angst. He claims that he reanalyzed their data and discovered that they actually *supported* his thesis. Sulloway reached this conclusion in part by rejecting certain birth-order studies in which subjects assessed their own personal-ities, including the enormous study that Ernst and Angst had carried out in Zurich. These "self-report" studies are virtually worthless, Sulloway contended, because peoples' views of themselves are hopelessly biased.

Reached in Zurich, Angst informed me by e-mail that he could "neither reconstruct nor understand" how the data gathered by him and Ernst for their 1983 book had been reanalyzed by Sulloway. Sulloway had produced "an impressive series of case histories and illustrations," Angst said. "But to my mind he is dealing in a different currency: historical investigation is by nature retrospective, unrepresentative and not generalizable and cannot disprove the findings of well-planned empirical investigations."

Sulloway's methods have also been faulted by psychologist Judith Harris. In her 1998 book, *The Nurture Assumption*, Harris noted that Sulloway reached his conclusions by relying heavily on studies in which subjects' personalities were assessed by relatives—namely, siblings and parents. This method is particularly ill suited for testing Sulloway's birth-order thesis, according to Harris. The essence of the thesis, she explained, is that the competitive strategies that siblings employ within their families during childhood carry over into their relationships outside the home and persist into adulthood. "Birth order effects are frequently found in ratings by parents and siblings; they are generally absent from measurements made outside the family context," Harris said.

Darwin and Freud

I heard a surprisingly similar complaint about *Born to Rebel* from Steven Pinker, Sulloway's fellow Darwinian and MIT colleague. Although he praised *Born to Rebel* in *How the Mind Works*, Pinker told me that he had doubts about Sulloway's thesis; it seemed too, well, *Freudian*. Like Freud, Sulloway assumes that "our way of interacting with our family determines how we interact with the outside world," Pinker explained. "That's not obviously true. I'm skeptical of it."

The irony of Pinker's remarks was striking. Before the publication of *Born to Rebel*, Frank Sulloway was known primarily as a critic of Freud. In his 1979 book, *Freud, Biologist of the Mind*, Sulloway argued that Freud, far from being an original thinker, borrowed freely from other scientists, ranging from Darwin and Lamarck to Wilhelm Fliess, a neurologist who was convinced that the nose was the crux of many psychological disorders.

Actually evolutionary psychologists have more in common with Freudians than they usually like to admit. Both view sex as the key to the human psyche. Both see men and women as fundamentally different and in certain respects incompatible. Both share an essentially tragic view of human nature: life is a struggle, and happiness is fleeting, if attainable at all. Both are acutely aware of the limits of our rationality and the power of our instincts. Both propound theories that are almost infinitely flexible, that can explain virtually every fact of human psychology and behavior and are thus immune to falsification.

The unconscious looms as large in evolutionary psychology as it does in psychoanalysis. Like the Freudians' id, the Darwinians' selfish genes motivate us in ways of which we are generally unaware. Like many other ideas in evolutionary psychology, this one can be traced back to a hypothesis advanced by Robert Trivers in the 1980s. He proposed that there are good evolutionary reasons for us to exaggerate our own trustworthiness, generosity, and strength and to downplay our selfishness, infidelity, and other shortcomings. The most effective liars, according to Trivers, are those who believe their own lies and thus project an air of sincerity. On the other hand, we should not be so self-deluded that we cannot learn from our mistakes. As George Orwell once said, "The secret of rulership is to combine a belief in one's own infallibility with a power to learn from past mistakes."

Like the Freudians, Trivers has also depicted the family as a cauldron of conflict. The divergent genetic interests of family members could generate tension not only between siblings but also between parents and their offspring. Inspired by these ideas, Margo Wilson and Martin Daly (discoverers of the evil-stepparent syndrome) have proposed a Darwinian version of the Oedipus complex. According to Wilson and Daly, boys and girls should be equally jealous of their parents' sexual relationship, which threatens to produce more offspring and thus reduce each child's share of the parental resources. The children thus demand so much attention that their parents are too exhausted to contemplate sex. The tension should be particularly pronounced between the children and the father, who is more motivated than the mother to keep producing more children—that is, to have more sex.

Finally, Darwinians, like Freudians, are not above employing their theoretical instruments to dissect their critics. The evolutionary psychologist David Buss has utilized this tactic. Buss charged in *Psychological Inquiry* that resistance to evolutionary psychology stems less from legitimate scientific concerns than from the desire of psychology's old guard to defend itself from a powerful upstart; "traditional psychologists" fear a loss in status and prestige (and, I would assume, sexual opportunities!). Buss suggested that research be done to determine whether critics of evolutionary psychology tend to be firstborns, who according to Frank Sulloway are constitutionally resistant to change.

Our Improbable Past, and Future

Evolutionary psychology is in many respects a strangely inconsequential exercise, especially given the evangelical fervor with which it is touted by adherents. Evolutionists can take any set of psychological and social data and show how they can be explained in Darwinian terms. But they cannot perform experiments that will establish that their view is right and the alternative view is wrong—or vice versa. This quandary reminds me of the field of physics devoted to "interpreting"—that is, determining the metaphysical meaning of—quantum mechanics. Many different interpretations have been proposed, including the Copenhagen interpretation, the many-worlds hypothesis, and the pilot-wave theory. The problem is that each interpretation explains the available data, and thus there is no way to determine empirically which interpretation is correct. One is forced to choose based on aesthetic preferences. In the same way, empirical data alone cannot determine where evolutionary psychology is right and cultural determinism is wrong. One is forced to fall back on aesthetic, political, or philosophical preferences.

Although I do not think it has proved to be terribly useful for understanding *human* nature, I tend to agree with the philosopher Daniel Dennett that Darwin's theory of evolution by natural selection is "the single best idea anyone has ever had." Dennett wrote in *Darwin's Dangerous Idea* that Darwin's insight "unified the realm of life, meaning and purpose with the realm of space and time, cause and effect, mechanism and physical law." Evolution through natural selection should no longer be called a theory; it is a fact, as well demonstrated as any other in science.

But that immediately raises a question. Why, if evolutionary theory is so well supported, must people like Daniel Dennett, Richard Dawkins, and others expend so much energy extolling its charms? Why does Darwinism, as a theory of all of nature (rather than just the human variety), stick in the craw not only of religious fundamentalists but also of many extremely knowledgeable scientists? Some critics, notably the left-leaning biologists Stephen Jay Gould and Richard Lewontin (and probably Noam Chomsky, in spite of his disavowals), clearly have political objections to Darwinian theories. They fear that if we accept adaptationist explanations of nature, we may come to believe that many unpleasant features of modern life—ruthless capitalism, racism, sexism, nationalism, and the like—were to some extent probable and even inevitable outcomes of evolution and not easily subject to change. Given how genetic theories have been employed in the past, such concerns are not unwarranted.

But others object to Darwinism for precisely the opposite reason. They fear that evolutionary theory, even when buttressed by modern genetics and molecular biology, does not make reality probable *enough*. Darwinism cannot tell us why life appeared in the first place or why, once it emerged, it took the course it did. Scientists have proposed various auxiliary mechanisms to make life appear more probable and robust, including group selection, Gaia, and complexity theory, but none are very plausible. The particle physicist Steven Weinberg once wrote, "The more the universe seems comprehensible, the more it seems pointless." The history of biology suggests a corollary aphorism: the more life seems comprehensible, the more it seems improbable. The most wildly improbable organism of all is the one that can fret over its improbability.

Evolutionary theory raises questions about our future as well as our past. Just how far can evolution go? Will humans keep getting smarter? Will *Homo sapiens* give rise to some other more intelligent species some day, just as apes gave rise to us? Not surprisingly, evolutionists cannot agree on the answer. Some argue that natural selection has been so attenuated by modern medicine and other products of civilization that it is unlikely to bring about any dramatic changes in the bodies or minds of humans. Barring tremendous breakthroughs in behavioral genetics, or the advent of widescale eugenics programs, we are, basically, what we will be.

Other theorists speculate that our descendants may undergo drastic changes. In the most extravagant scenario, the next stage in evolution will belong to machines whose intelligence dwarfs that of any mere human. Even during Darwin's era, such fantasies were so common that Samuel Butler satirized them in his science-fiction novel *Erewhon*. A mad-scientist character in the 1872 novel proclaimed, "There is no security against the ultimate development of mechanical consciousness, in the fact of machines possessing little consciousness now. . . . The more highly organized machines are creatures not so much of yesterday, as of the last five minutes, so to speak, in comparison with past time. Assume for the sake of argument that conscious beings have existed for some twenty million years: see what strides machines have made in the last thousand! May not the world last twenty million years longer? If so, what will they not in the end become?"

This same rhetoric, minus the irony, is still employed by artificial intelligence enthusiasts such as Marvin Minsky of the Massachusetts Institute of Technology and Hans Moravec of Carnegie

Mellon University. They have envisioned a day when not just the earth but the entire universe will be ruled by superintelligent machines. Some neo-Darwinians roll their eyes. "Why are there so many robots in fiction, but none in real life?" Steven Pinker asked in *How the Mind Works*. "I would pay a lot for a robot that could put away the dishes or run simple errands. But I will not have the opportunity in this century, and probably not in the next one either." The reason, Pinker explained, is that "the engineering problems that we humans solve as we see and walk and plan and make it through the day are far more challenging than landing on the moon or sequencing the human genome." In other words, HAL from *2001* and R2D2 from *Star Wars* may remain forever in the realm of science fiction.

Actually evolutionary psychology and artificial intelligence share a common problem in describing human mental life, according to the philosopher and cognitive scientist Jerry Fodor of Rutgers University. Fodor was once a leading proponent of the computational theory of mind—and of the notion that the mind is not an all-purpose learning device but is divided into modules dedicated to specific tasks. By 1998, however, Fodor was acknowledging and even dwelling on the limits of these assumptions. Certain cognitive tasks, such as the ability to detect color or to parse a sentence, can indeed be reduced to computation, Fodor noted in a review of *How the Mind Works*; but dividing the mind into many little dedicated computers, or modules, still leaves unanswered the question of how the results of all these modular computations become integrated. A psychology that "hasn't faced the integration problem," Fodor concluded, "has barely got off the ground."

To illustrate this problem, Fodor pointed out that design peculiarities of our vision module create various illusions; for example, the moon appears larger when it is near the horizon than it does overhead. But most of us know that we are seeing an illusion; the output of our vision module must somehow be integrated with the output of other modules to produce what might be called "common sense." "As things now stand," Fodor stated, "we don't have a theory of the psychology of common sense that would survive scrutiny by an intelligent five-year-old. Similarly, common sense is egregiously what the computers that we know how to build don't have." Like neuroscientists, researchers in evolutionary psychology and artificial intelligence are both bumping up against the Humpty Dumpty dilemma. They can break the mind into pieces, but they have no idea how to put it back together again. A crucial missing ingredient is plain old common sense.

7

ARTIFICIAL COMMON SENSE

....

My expert systems couldn't be called intelligent. But they did get me thinking about what could be. I thought about the question for a long time, even after I jettisoned the commercial interests.... After great inference, I came to the conclusion that I hadn't the foggiest notion what cognition was.

—RICHARD POWERS, *GALATEA* 2.2

In 1982 I took a class in science writing at Columbia University with the author Pamela McCorduck. She had already written one book about artificial intelligence, *Machines Who Think* (note the mischievous *Who*), and was hard at work on another. Her enthusiasm for artificial intelligence spilled over into her class. She regaled us with stories about Herbert Simon, Marvin Minsky, John McCarthy, and other pioneers of artificial intelligence (AI). Her new book was about Japan's Fifth Generation Project, which was intended to produce intelligent machines within a decade. McCorduck, who is married to the prominent computer scientist Joseph Traub, teased the would-be writers in her class with the notion that one day computers might write prose as well as, or better than, any humans. In *Machines Who Think* she had presented an even more apocalyptic scenario: according to some AI visionaries, intelligent machines represented the next phase in the evolution of conscious life and would soon leave mere humans far behind.

A year later I gained an only slightly more sober perspective on the AI culture when I took a job at *IEEE Spectrum*, the monthly journal of the Institute of Electrical and Electronics Engineers. With a membership exceeding 300,000, the IEEE is one of the largest professional societies in the world. AI was a major focus of *Spectrum*'s coverage. The U.S. Department of Defense was funneling vast sums into the field, as were industrial giants such as IBM. In 1984, to celebrate the hundreth anniversary of the IEEE, the editors of *Spectrum* put together a special issue on "the impact of high technology on both the citizen at large and the engineering professional." Titled "Beyond 1984: Technology and the Individual," the cover showed the head and shoulders of a silver, faceless robot beside a bald, almost equally expressionless woman. The message was clear: the breakneck pace of technological progress was blurring the divide between the artificial and biological realms.

The lead article, which I edited, had an innocuous title, "The Machine as Partner of the New Professional," with a provocative subtitle: "Today's instruments are providing more data than the mind can process; expert systems are coming to the rescue but may soon usurp human roles." (Expert systems are software programs that mimic the ability of human experts to absorb information and make decisions based on it.) The author, Frederick Hayes-Roth, was a vice president of Teknowledge, a producer of expert systems, and a former director of the AI program at the Rand Corporation.

Needless to say, he was an evangelist for artificial intelligence. In his *Spectrum* article, he predicted that electrotechnology would "dramatically alter the nature of professional work as we know it. Some professions will be made obsolete and disappear. Others will change substantially. Many new professions will emerge to exploit and support new applications of electronics. The most profound challenge for future professionals, however, will arise when machines begin to assume their intellectual responsibilities."

Hayes-Roth prophesied that expert systems would supplant air traffic controllers by the year 2000 and doctors and scientists within as few as fifty years. As automated expertise surpasses the human variety, experts will lose cultural prestige; creativity, athletic prowess, empathy, and intuition, which are more difficult to automate, will increase in value. Humans will "pursue interests that are almost exclusively personal and social rather than economic, due to the takeover by machines of most income-producing tasks. Indeed, if peace prevails, the greatest tasks for humanity may be seeking out new challenges and redefining what it is to be human." One niche in which humans would thrive was psychotherapy. "Professionals dealing with human problems—traumas of adolescence, adjustment to life in outer space, divorce, aging and death—will expand in number," Hayes-Roth declared.

In 1998, fourteen years after I had last spoken to Hayes-Roth, I tracked him down in Palo Alto, California. He was still working for Teknowledge, the expert system company that he had helped to found in the glory days of AI. After reminding him that I had edited his article for *Spectrum* in 1984, I said that I had called to ask how he thought his predictions had held up. He started laughing. "You've got a real mean streak," he said. He cheerfully admitted that since the early 1980s the field of expert systems, and AI generally, had stalled.

Expert system designers had run into two problems. Extracting knowledge from human experts and turning it into software that could address real problems turned out to be an extremely arduous, time-consuming task. The knowledge accumulated for one project usually had little or no relevance for the next. "It is an example of what is called nonrecurrent engineering," Hayes-Roth said. AI pioneers also underestimated how difficult it would be to mimic human cognition in all its complexity, Hayes-Roth remarked. The kind of reductionism that works in physics or molecular biology fails when it comes to the human mind. Humans are "very, very complicated systems that are both evolved and adapted through learning to deal well and differentially with dozens of variables at one time."

Hayes-Roth remained an optimist, however. The major obstacles to AI, he said, were political and economic rather than scientific. For example, the Federal Aviation Administration could have automated its antiquated air traffic control system by the year 2000, as Hayes-Roth had predicted, if not for bureaucratic inertia. Many promising AI projects had been terminated because their funders, governmental or industrial, had become too impatient. In spite of these problems, Hayes-Roth said, AI researchers had accomplished a great deal, such as speech-recognition devices and software programs that can write other software programs. During the Persian Gulf War in 1991 the U.S. Air Force used an expert system to plan air sorties against Iraq. "The air force wrote that this expert system more than justified" the Defense Department's entire investment in AI, Hayes-Roth said.

Hayes-Roth still had faith that the dream of building a truly intelligent machine would be fulfilled. "It's really just a matter of time and financing and organization," he said. He saw no reason that engineers could not build a computer like HAL, the lip-reading cyber-villain of the movie *2001*. "I know as well as anybody what are the barriers," Hayes-Roth said. "On the other hand, if somebody were to assemble a team, like a Manhattan Project, to build HAL, that team would succeed. It would take on the order of a decade or maybe two, but it's not going to be an infinite effort."

Herbert Simon's Prophecies

Hayes-Roth is a craven defeatist compared to Herbert Simon of Carnegie Mellon University, a founding father of both AI and cognitive science. Unlike many other AIers, Simon refuses to acknowledge that AI has in any sense failed to live up to expectations. When I asked him why the enormous surge in machine intelligence foreseen back in 1984 by Hayes-Roth and others had not happened, Simon retorted, "It *has* happened. Not on that particular scale and direction, but it has happened." Simon granted that building intelligent machines had been in some respects more difficult than anticipated. "AI turned out to have a relatively easy part and a relatively hard part," Simon said. "The relatively easy part was how people think deep thoughts." AIers have made great progress in simulating the human ability to form and manipulate symbolic representations of the world, Simon claimed.

The hard part was simulating how people interact with the environment through their senses and motor organs. "Interface with the environment, that's really the complicated thing." Simply keeping track of changes in the environment was extremely difficult, especially given the limited power of computers in the early days of AI. "It wasn't until maybe ten or fifteen years ago that we had computers fast enough and big enough to deal, for example, with dynamic images instead of static images." Researchers have also invented tricks to speed up image processing, such as a program that rapidly detects the motion of objects by subtracting each current image from the preceding one.

Robots and computers are "still very far from the versatility of humans in doing these particular things," Simon said. But that situation is changing fast, he added. He rattled off various projects that demonstrate how far AI has come lately. One is a car, called Navlab, that has driven across the country with minimal help from a human. Human intervention was required mainly "getting on and off superhighway ramps," Simon said. Robots deliver the mail at a hospital in Pittsburgh and elsewhere. Simon was particularly excited by tournaments that pit teams of soccer-playing robots against each other. Each robot had to cooperate with its teammates and to anticipate and thwart the plans of its opponents.

Asked if he considered AI to be primarily science or engineering, Simon replied, "It's both. It's a study of the engineering of intelligence, or the science of intelligence, if you like, with no holds barred." Ideally, Simon said, AI models should mimic not only the human mind's output but also its internal processes. Determining whether a computer model faithfully mirrors the brain's functioning is still difficult, because magnetic resonance imaging and other technologies cannot reveal neural processes in sufficient detail. "I think it'll be a little while before we have that bridge, but we don't have to get nervous breakdowns about that." Advances in computers on the one hand and in brain imaging on the other would soon bridge the gap between AI and neuroscience.

Simon is a firm believer in what is sometimes called the strong AI program, which holds that there is no fundamental difference between computers and brains. "A computer is different machinery than a person in terms of speeds and memory capacity, et cetera," Simon explained. "I think almost *only* in terms of those two." Another tenet of strong AI is that a machine performing some cognitive task, such as playing chess or recognizing a face or playing soccer, is thinking by definition.

According to Simon, most of the criticism of AI has been motivated by emotion rather than a cool appraisal of the facts. "This field gets judged by a different standard than any other field I know of." Intelligent machines represent a severe blow to humanity's vanity, Simon said. "Most people find just repulsive the idea of a computer thinking." Every time he gives a talk on AI, Simon informed me, questions from the audience reflect people's "fear and concern" that they will be replaced by machines.

This complaint is common among AI researchers. In 1998, the *New York Times* published a defense of AI by Astro Teller, who like Simon is an AI researcher at Carnegie Mellon. When we express skepticism toward AI, Teller suggested, we are really in denial. Intelligent computers make us feel less special; we feel threatened by them, just as we once felt threatened by the discovery that the earth is not the center of the universe and that humans descended from apes. As a result, Teller asserted, critics keep "raising the bar as artificial intelligence makes progress, so that they don't have to admit that machines can be creative or intelligent."

Actually, if anyone has doomed AI to perpetual failure by setting the bar impossibly high, it is AI enthusiasts like Simon. In 1957 he spoke of AI in almost mystical terms. Artificial intelligence "will help man obey the ancient injunction: Know thyself. And knowing himself, he may learn to use advances of knowledge to benefit, rather than destroy, the human species." In that same lecture Simon made four slightly more specific predictions. He prophesied that within ten years—in other words, by 1967—AI would reach the following milestones:

- A computer will be the world's chess champion.
- A computer will discover and prove an important new mathematical theorem.
- A computer will write music that will be accepted by critics as possessing considerable aesthetic value.
- Most theories in psychology will take the form of computer programs, or of qualitative statements about the characteristics of computer programs.

Simon assured me in 1998 that his prophecies had held up quite well. The chess prediction had been off by thirty years, but the other three "were right on the nose." But to what extent were Simon's forecasts really fulfilled? Simon's safest bet was artificial music composition. At the time of Simon's speech in 1957, the computer scientists L. A. Hiller and L. M. Isaacson had already written a program that generated music. Simon himself described one of the program's compositions, the *Illiac Suite* (named after one of the first digital computers), as "not trivial and uninteresting." Computer-generated music went on to become a minor fad, just as computer-generated art and poetry did. The melodies, drawings, and poems produced by computers often elicit strong reactions from humans. But as one journalist noted in a story on artificial music and art in 1997, "People also see images of Jesus in burnt tortillas and attribute feelings to their cars." Not even the most avid AI enthusiast claims that these programs represent a serious challenge to human composers, artists, and poets.

As for artificial mathematics, various computer scientists, including Simon himself, did indeed develop programs that posed and even solved mathematical theorems. In the late 1950s Herbert Gelernter, a physicist at IBM, created a program that "discovered" some of Euclid's basic theorems of geometry. Other computer scientists invented algorithms that could churn out more advanced theorems. But even within the AI community, critics complained that the results achieved by these programs had been embedded in them from the start. Since the late 1970s mathematicians have become increasingly reliant on computers to perform enormous calculations needed to complete certain proofs. But the computer is not constructing the proof; it is merely serving as a tool for its human master.

In the early 1980s the computer scientist and entrepreneur Edward Fredkin sought to revive the flagging interest in artificial mathematics by creating what came to be known as the Leibniz Prize. (Fredkin was one of the AIers who had suggested to Pamela McCorduck in *Machines Who Think* that computers represent the next stage in the evolution of intelligence.) Administered by Carnegie Mellon, the prize offered $100,000 for the first computer program to devise a theorem that has a "profound effect" on mathematics. One of the original judges for the prize was David Mumford, a mathematician at Brown University. I once asked Mumford when he expected the prize to be claimed. "Not now, not 100 years from now," he replied.

Simon's chess prediction probably seemed relatively modest to many observers. Chess is based on simple, unambiguous rules and is played on a tiny cartesian playing field. In principle, a computer could calculate the consequences of each possible move, examining all possible counter-moves, counter-counter moves, and so on until it finds a winning strategy. In practice, that strategy is impossible. At any given moment, a chess player typically has thirty-eight different possible moves. Each of those possible moves generates thirty-eight possible countermoves by the other player, or 1,444 moves in all. A chess player trying to see just two moves and countermoves ahead has 2,085,135 possibilities to contemplate. The total number of games that the rules of chess can generate is 10^{120}, or more than the number of atoms in the universe.

Engineers did not succeed in designing a machine that could carry out the basic rules of chess until 1958. After that, chess-playing machines made slow but steady progress. By the mid-

1960s, computers could play as well as average tournament players, and they steadily began climbing up the rankings. Although it took thirty years more than he had predicted, Simon was, needless to say, delighted when the IBM computer Deep Blue finally defeated chess champion Gary Kasparov in 1997. Deep Blue was a prodigiously powerful machine. Its thirty-two separate microprocessors were capable of examining 200 million chess positions per second. With this power, plus some rules for excluding unpromising lines of attack, Deep Blue could look thirty-five moves into the future.

Ironically, Deep Blue's own handlers have questioned whether their success represents a vindication for artificial intelligence. "This chess project is not AI," said Chung-jen Tan, manager of the Deep Blue team, when I interviewed him and his colleagues in 1996, just after their first match with Kasparov. (In that contest, Deep Blue won the first game but eventually lost the match.) Deep Blue succeeds not by trying to mimic human judgment but by reducing the problem to pure computation, Tan said. "Before you understand the problem, you call it AI," he elaborated. "But once you really understand it, you can reduce that to a computation algorithm." "The techniques that tried to mimic human judgment failed miserably," Tan's colleague Joseph Hoane concurred. "We still don't know how to do that at all."

Deep Blue's team seemed amused rather than flattered by Herbert Simon's claim that Deep Blue could "think." ("I would call what Deep Blue does thinking," Simon had declared to the *New York Times* in 1996.) "I don't think chess playing has any bearing on whether computers can think," IBM scientist Murray Campbell said brusquely. "Just because a computer can play chess doesn't mean it can think. 'Thinking' is a very difficult word." The Deep Blue team also seemed to doubt whether engineers would be able to build a truly human-like computer any time soon, as Simon and other AI pioneers had once hoped. Campbell noted that one could build an artificial brain by substituting an electronic switching device for every neuron. "But that's, like, centuries in the future," Campbell said. "It's more than that," Tan added emphatically. "The brain is more than hardware. It's all the software and everything else. I'm not a psychologist or a neuroscientist, but I'm sure they don't understand those problems either."

"He's full of crap," Simon responded when I mentioned Tan's comment that Deep Blue "is not AI." If Deep Blue was relying on number crunching alone, Simon asked, why did the IBM team employ grandmasters to fine-tune the program? The Deep Blue team, Simon contended, was merely parroting the IBM party line. Since the late 1950s, he elaborated, IBM has discouraged its employees from associating its products with AI or even using the term. "IBM is scared out of its wits that somebody is going to think they are delivering thinking machines" that will replace human workers, Simon said.

Perhaps the most significant and prescient of Simon's prophecies was that computer programs would play an increasingly important role in psychology. In 1957, when Simon made his prediction, the dominant school of psychology was behaviorism, which treats the mind as a black box, that is, an object whose internal structure is unknown and even irrelevant and thus can be ignored. But behaviorism soon gave way to cognitive science (also called cognitive psychology or cognitive neuroscience), which views the mind as an information-processing device and seeks to unravel the computational underpinnings of pattern recognition and other components of cognition.

The ascent of cognitive science has indeed spurred the proliferation of computational models of human cognition. These models fall into two broad categories: rule-based algorithms and neural networks. With the rule-based method, knowledge is embedded in the model from the start, typically in the form of if-then instructions that attempt to anticipate every situation. (If the object in your field of vision has orange and black stripes and a long tail and big teeth, then run in the opposite direction as fast as you can.)

Neural networks typically consist of several levels of nodes, or "neurons," linked by connections, or "synapses," of varying strength. Data are fed into the first level of neurons; if the incoming signal is sufficiently strong, the neuron "fires," sending a signal to all the neurons in the next level of the network. Each of those neurons, in turn, if it receives sufficient stimulation, transmits a signal to the next level of the network. When one neuron repeatedly signals to another, the connection between them is strengthened, making it easier for signal propagation to occur in the future.

As described by some journalists and even scientists, neural networks seem to possess almost mystical powers. Actually neural networks are a new-fangled embodiment of old-fashioned statistical methods, such as curve interpolation, for coping with incomplete or ambiguous data. The process, grossly simplified, works as follows: Input such as a two-dimensional image of a face is transformed into a set of points, or coordinates, each of which has an x and y value. Fed these data, the neural network searches for a curve, or mathematical function, that matches the coordinates as closely as possible. Each of these functions corresponds to a specific output—"tiger" or "house cat." Through various feedback mechanisms, the network can be "trained" to recognize patterns with greater accuracy.

One reason that many researchers favor neural networks is that, superficially at least, they resemble human brains. Unlike rule-based algorithms, which are front-loaded with all their knowledge, neural networks acquire knowledge—at least in principle—through a process roughly analogous to human learning. Neural networks also perform calculations not serially (one at a time) but in parallel. But in practice, neural networks have turned out to be as limited and inflexible as rule-based methods. Knowledge must still be applied from the outside during and even beyond the training period.

The Philosopher's Revenge

Hubert Dreyfus doubts whether either neural networks or the rule-based approach will lead to truly intelligent machines. Dreyfus, a philosopher at the University of California at Berkeley, is one of the earliest and most persistent critics of AI. He turned his attention to AI while he was at MIT in the late 1950s and early 1960s. AI researchers were claiming that philosophers "had wasted 2,000 years trying to understand things like perception and memory and language and thinking and so forth, and [AI researchers] had taken over and were making great progress," Dreyfus recalled. These claims intrigued Dreyfus. Unlike many other philosophers, he had little interest in the metaphysical conundrum of whether computers would ever be conscious; his focus was on what computers could do. Could they solve important problems in science or mathematics? Translate Goethe into English? What about holding a conversation or recognizing a familiar face in a crowded room?

Dreyfus was dubious. He first spelled out his doubts in a paper published by the Rand Corporation, a major think tank, in 1967. The paper was expanded into *What Computers Can't Do*, originally released in 1972. Dreyfus wisely did not claim to have proved that AI is impossible; in fact, he questioned whether such an impossibility proof could be constructed. But he presented various reasons that AI might not progress as fast as proponents hoped. At the time, the major trend within AI—and within the burgeoning and overlapping field of cognitive science—was to view cognition as a rule-based process. According to this view, when we recognize a celebrity on television or remember where we left our keys, our brains are carrying out multitudes of if-then instructions, albeit below the level of awareness.

Dreyfus cited various philosophers, notably Wittgenstein and Heidegger, who had argued that it would be extremely difficult to reproduce human perception and cognition with a formal, rule-based model. Wittgenstein asserted that any factual statement about the world had to be explained by other factual statements; there are no fundamental facts, or "primitives," as Wittgenstein referred to them, that serve as the foundation for cognition, as quarks and electrons do for the physical realm. Similarly, Heidegger pointed out that rules rarely apply to all situations; each rule requires extra rules to determine whether the initial rule is relevant for a particular situation, thus creating an infinite regress of rules.

AI theorists could and did argue that one could surmount these obstacles by finding the elusive cognitive "primitives" or by creating sufficiently clever rules, or axioms; after all, the human brain somehow accomplishes this trick. As Dreyfus explained, "You could always say, 'Well, we'll just get such good relevance axioms and such good meta-axioms for cutting down the relevance axioms that it'll work.'" But reducing a cognitive task to a rule-based procedure is in practice extraordinarily difficult. Even if a cognitive task *can* be defined in terms of rules, Dreyfus argued, it may still be difficult for a computer to carry out. For example, human chess experts rely not on brute calculation but on their memory of what has worked in the past, on rules of thumb about which strategies are best, and on what might be called intuition. The reason that chess is such a compelling game is that this expertise cannot easily be reduced to rules.

As difficult as it is to write down the rules for playing grandmaster chess, Dreyfus said, it should be exponentially more difficult to find a set of rules that can tell a computer how to carry on an ordinary conversation. The rules of grammar and syntax are vast, complex, and riddled with exceptions. A single word may have many different meanings and connotations, which vary depending not only on the sentence within which the word appears but also on the context of the conversation.

What is most striking about human intelligence, Dreyfus pointed out, is its capacity for rapidly processing and acting on ambiguous, open-ended data. Our ability to understand a sentence—or to recognize a face or walk down a crowded city street—depends to a large degree on our ability to draw on a vast reservoir of worldly knowledge that might be called common sense. "Is an exhaustive analysis of human reason into rule-governed operations on discrete, determinate, context-free elements possible?" Dreyfus asked toward the end of *What Computers*

Can't Do. "Is an approximation to this goal of artificial intelligence even probable? The answer to both these questions seems to be no."

Dreyfus became the token critic of AI at conferences and in popular articles. "When anyone wrote a gung-ho article about the marvels of computers," he recalled, "they always had a paragraph or two on me and my views, which they then proceeded to ignore for the rest of the article." The AI community did score at least one significant public relations victory against Dreyfus. Dreyfus reveled in pointing out how far chess programs fell short of their makers' goals; he gleefully reminded AIers that in 1960 a ten-year-old boy defeated a machine Herbert Simon designed. In 1966 Dreyfus accepted a challenge to play another chess-playing computer, called MacHack. The machine won. "Computers Can't Play Chess. Nor Can Dreyfus," a computer science newsletter reported merrily.

Dreyfus denied that he had predicted computers would *never* play better than a ten-year-old boy, as some AIers have suggested. He thought computers would eventually excel at chess, because chess, more than other cognitive skills, can be reduced to formal rules. But he admitted that he had not expected Deep Blue to conquer Gary Kasparov in 1997. Moreover, he was "very surprised" at the growing ability of computers to recognize spoken words. "It takes a lot of speed and a lot of [memory capacity], but it can be done," he said. On the other hand, computers still cannot talk about politics or understand a fairy tale; they still lack the ordinary skills that "enable us to find our way around in the world and recognize what's relevant." In short, they lack common sense.

In 1992 MIT Press released a new edition of *What Computers Can't Do*, retitled *What Computers Still Can't Do*. In the introduction, Dreyfus proclaimed victory over what he called "good old-fashioned AI." After fifty years of effort, "it is now clear to all but a few diehards that this attempt to produce general intelligence has failed." Dreyfus was pleased that many AIers had abandoned the rule-based method for neural networks, which struck him as a more plausible model of human cognition. But he noted that in practice, neural networks bump up against the same problem as good old-fashioned AI: an inability to duplicate common sense. "One needs a learning device that shares enough human concerns and human structure to learn to generalize the way human beings do." Borrowing a term from the philosopher of science Imre Lakatos, Dreyfus called AI a "degenerating research program." Such a program

> starts out with great promise, offering an approach that leads to impressive results in a limited domain. Almost inevitably researchers will want to try to apply the approach more broadly, starting with problems that are in some way similar to the original ones. As long as it succeeds, the research program expands and attracts followers. If, however, researchers start encountering unexpected but important phenomena that consistently resist the new techniques, the program will stagnate, and researchers will abandon it as a progressive alternative approach becomes available.

The failure of AI to mimic the mind, Dreyfus pointed out to me, mirrors the larger failure of psychology to comprehend the mind. Over the past century, psychology has failed to produce a single framework or paradigm powerful enough to command the assent of most researchers. "It's a bunch of little paradigms, each one of them saying the other one is totally hopeless," he said. "The behaviorists look like they have the right answer, and then Chomsky writes one review article about Skinner, and that bashed behaviorism. And then it looks like rules and cognitivism are the right answer. And now [neural networks are] coming along. . . . It's just a bunch of *fads*. It doesn't seem to get anywhere. There's been no progress in understanding the mind."

Douglas Lenat's Assault on Common Sense

Many former believers in the computational theory of mind have reluctantly accepted Dreyfus's judgment, although, needless to say, they rarely credit Dreyfus. That message came through in *HAL's Legacy*, a collection of essays by leading AIers published in 1997. (1997 was the year that HAL "became operational" at a plant in Urbana, Illinois, in the novel *2001*; in the movie *2001*, HAL's birth year was 1992.) "Let's state the obvious: HAL doesn't exist, and there is no chance that some miraculous change in funding or insight will yield AI at the level portrayed in HAL by the year 2001," declared David Stork, the book's editor, in an introductory chapter. Stork, a computer scientist associated with Stanford University and the Ricoh California Research Center, noted that he had created moderately successful lip-reading programs, but "no current system even remotely approaches HAL's proficiency at speechreading in silence."

"As we approach 2001," Stork continued, "we might ask why we have not matched the dream of making a HAL. The reasons are instructive. In broad overview, we have met, and surpassed, the vision of HAL in those domains—speech, hardware, planning, chess—that can be narrowly defined and easily specified. But in domains such as language understanding and common sense, which are basically limitless in their possibilities and hard to specify, we fall far short."

"Under any general definition . . . AI so far has been a failure," agreed veteran computer scientist David Kuck. Roger Shank of Northwestern University declared flatly that HAL "is an unrealistic conception of an intelligent machine" and "could never exist." The best that computer scientists can hope to do is to create machines "that will know a great deal about what they are supposed to know about and miserably little about anything else."

This pessimism was vehemently rejected by one contributor to *HAL's Legacy*, Douglas Lenat. Lenat accepted the consensus of other contributors that common sense was the key to AI's success—and failure. "We're now in a position to specify the steps required to bring a HAL-like being into existence," Lenat declared:

1. Prime the pump with millions of everyday terms, concepts, facts, and rules of thumb that comprise human consensus reality—that is, common sense.
2. On top of this base, construct the ability to communicate in a natural language, such as English. Let the HAL-to-be use that ability to vastly enlarge its knowledge base.
3. Eventually, as it reaches the frontier of human knowledge in some area, there will be no one left to talk to about it, so it will need to perform some experiments to make further headway in that area.

Lenat revealed that "this is not just a fanciful blueprint for some massive future endeavor to be launched when humanity reaches a higher plateau of utopian cooperation. It is, in fact, the specific plan I and my team have been following for the past dozen years."

Lenat is one of the few AI researchers in the world who has tried to sustain the original AI vision of building a computer with an all-purpose rather than highly specialized intelligence. As early as 1984, he had concluded that specialized programs represented a dead end for AI. All of them hit "the very same brick wall— namely the need for our programs to have the same breadth and depth of common-sense knowledge as people do." That same year, Lenat took on the challenge of constructing a computer program that contained the same kind of knowledge that virtually every human has. Called Cyc, it was funded initially by a consortium of high-technology firms called MCC (for Microelectronics and Computer Consortium), based in Austin, Texas. Lenat left MCC and formed his own company, Cycorp, in 1994.

Lenat and his researchers could not simply extract knowledge from dictionaries and encyclopedias. Even the simplest encyclopedia entry, they pointed out, contains deep implicit assumptions. Consider the following sentences: "Napoleon died on Saint

Helena. Wellington was greatly saddened." The reader is supposed to know implicitly that Saint Helena is a place, Napoleon and Wellington are people, Wellington and Napoleon are both dead, Wellington outlived Napoleon, and so on. Lenat and his programmers, whom he sometimes calls "ontologizers," have now compiled more than one million commonsense rules, or assertions. Cyc "knows" that trees are usually outdoors, that once people die they stay dead, and that a glass filled with milk will be right-side-up, not upside-down.

Lenat has made some even grander claims for Cyc. He contended in one interview that Cyc is already "self-aware." "If you ask it what it is," he explained, "it knows that it is a computer. If you ask who we are, it knows that we are users. It knows that it is running on a certain machine at a certain place in a certain time. It knows who is talking to it. It knows that a conversation or a run of an application is happening. It has the same kind of time sense that you and I do."

Yet in *HAL's Legacy* Lenat denied that Cyc could ever have any "feelings." "HAL, Cyc, and their ilk won't have emotions," he explained, "because they are not useful for integrating information, making decisions based on that information, and so on. A computer may pretend to have emotions, as part of what makes for a pleasing user interface, but it would be as foolish to consider such simulated emotions real as to think that the internal reasoning of a computer is carried out in English just because the input/output interface uses English."

In downplaying the importance of emotion, Lenat was bucking one of the latest trends in mind-related science. A growing number of cognitive scientists, AI researchers, and neuroscientists (notably Joseph LeDoux, whose work I discussed in Chapter 1) have insisted that emotion is crucial to human cognition and creativity. In fact, Lenat is by his own admission less a scientist, a seeker of truth, than an engineer, a builder of things. The goal of the Cyc team, he stated, "was not to understand more about how the human mind works, nor to test some particular theory of intelligence. Instead, we built nothing more nor less than an artifact, taking a very nuts-and-bolts engineering approach to the whole project."

So far, the commercial applications of Cyc are hardly earth shattering. Lenat has emphasized Cyc's usefulness as a search engine for extracting information from the World Wide Web or other databases. Most search engines seek exact matches of key words or phrases, or perhaps synonyms stored in a thesaurus. But Cyc could find matches based on the broader meaning of a word, phrase, or sentence. For example, an art director could use Cyc to find images corresponding to such loose criteria as "a happy person." By checking the captions of images in a database, Cyc would be able to refer the art director to a photograph of "a man watching his daughter taking her first step." Similarly, a request for an image of "a strong and adventurous person" might turn up a photograph of "a man climbing a rock face." Cyc cannot fulfill requests by scanning images directly, as a human photo researcher would; it must rely on matches between its request and the verbal descriptions attached to them.

Cyc could also detect the kind of errors and inconsistencies that ordinary computers never notice, according to Lenat. For example, financial databases often include information about the sex of an individual and his or her spouse. "Without having to be specially programmed for the task, Cyc would know that there's probably a mistake in the data if X and X's spouse have the same gender," Lenat said. (Lenat neglected to mention that some jurisdictions, such as San Francisco, allow same-sex marriages.)

Ironically, interacting with this "commonsense machine" requires highly specialized knowledge that in no way resembles common sense. Cyc does have a limited ability to interpret and act on instructions in ordinary English. But for the most part it can only receive knowledge translated into a complex system of logic known as second-order predicate calculus. It generally yields information in this same format.

Lenat's goal has always been for Cyc to become intelligent enough to acquire new knowledge on its own by scanning newspapers, books, and other sources of information. Lenat keeps pushing back the date when Cyc will achieve this capability. When he started the project in 1984, Lenat predicted that Cyc would become self-educating within ten years. In 1991, Lenat was still hopeful that by 1994 or 1995 Cyc would be able, as one reporter put it, to "pick up new knowledge more readily by reading than by having knowledge engineers spoonfeed it." In 1997, Lenat had pushed the date to 2001. By that date, Lenat said, Cyc would be capable not only of learning but of creating. Cyc will become a "full-fledged creative member of a group that comes up with new discoveries. Surprising discoveries. Way out of boxes."

Rodney Brooks Seeks the Elixir of Life

Hubert Dreyfus admired Lenat for trying to fulfill the original goal of AI: creating a machine with general-purpose skills. "I do respect him for it," Dreyfus remarked. But so far Lenat has fallen far short of his most important goal: Cyc cannot acquire knowledge on its own by reading newspapers and other publications. Lenat "has managed to make a CD with a common-sense encyclopedia on it," Dreyfus said.

The AI researcher Rodney Brooks offered a similar assessment of Lenat: "I love Doug. I admire his ambition. It's just totally wrong." Brooks called Cyc "a convenient makework project" that allowed Lenat "to avoid having to think." Cyc is essentially a thesaurus, Brooks said, that never really engages the world directly through any kind of sensory apparatus. "Ultimately you have to ground it out. You have to attach it to some other sensory-motor experience, and that's what I think he's missing." Lenat's version of common sense in no way resembles the human variety. "You still just have this dictionary up there, and that's not how people work."

Brooks, who was born and raised in Australia and still speaks with a faint Down Under twang, is one of the more intriguing figures in AI. He evolved from being a gadfly of the field to a pillar of the establishment; in 1997, he was named director of MIT's Artificial Intelligence Laboratory. He is an amiably cantankerous man with longish, frizzy hair and eyes that bulge when he becomes enthused (which is often). His charm comes from the fact that he deprecates himself along with everyone else. Now that he was directing a prestigious laboratory with a large budget and two hundred employees, he told me shortly after his promotion, he was trying to curb his rebelliousness. "I have to be an old, stodgy fuddy-duddy these days," he said. "In the mid- to late '80s I went around telling everyone what they'd done was wrong, and they got upset. I still believe that," he added, "but now they believe that I don't really believe that, so they're not so upset with me."

Like Herbert Simon, Brooks was irritated by much of the criticism of AI. "It's become fashionable to say AI was a failure, but I don't think it was as big a failure as people make out." Tax return programs, computer-enhanced kitchen appliances, computer games, and a host of other commercial products are all embodiments of the AI vision. But Brooks acknowledged that the goal of building truly intelligent machines had not been achieved. "We haven't built HAL, we haven't built Commander Data," he said, referring to the android character in the television series *Star Trek*. "I think we were going about it in the wrong way," Brooks said. "And now," he added with self-mocking grandiosity, sticking his finger into the air like a preacher, "I know the *right* way, and I'm going to make it *happen!*"

Brooks started out as a conventional AI researcher. For his Ph.D. thesis, he designed an extremely complex rule-based program for three-dimensional vision. He suffered a crisis of faith when he tried to build robots based on this program and found that they worked poorly, if at all. In trying to solve these problems, he made his program still more complex and cumbersome. Brooks also became annoyed by the claims of Herbert Gelernter, Herbert Simon, and others that they had created computer programs that could discover laws of physics and mathematical theorems. Like other skeptics, Brooks complained that such programs "discovered" only what their inventors had already embedded in them. "Even then, when I was a traditional AI person, I just couldn't stomach that. It made me so angry."

Brooks eventually rejected the whole approach to AI represented by these efforts. "Gradually it dawned on me," Brooks said, that just because a behavior can be described as deriving from a complex set of rules does not mean that is how it occurred. Beginning in the 1980s Brooks wrote a series of papers—with titles like "Elephants Don't Play Chess" and "Intelligence Without Representation"—challenging the notion that reason and logic are the key to intelligence. The problem with the reasoning-based approach to AI, Brooks said, is that it does not reflect how humans actually solve problems in the course of everyday life. "Humans are capable of going through logical chains of reasoning, but mostly it's post hoc rationalization." When asked to justify our actions, we "*make stuff up*," Brooks said, emphasizing each word.

No one suggests that insects employ reason and logic when they make decisions, and yet such animals display apparently sophisticated, complex behavior and problem-solving ability. Brooks decided that the complexity of biological behavior derives not solely from organisms themselves but from their interaction with a complex environment. He demonstrated his ideas by building dozens of insect-like robots whose behavior was governed by relatively simple sensors and chips programmed with simple rules. One basic rule instructed the insectoid, when it bumped into an obstacle, to keep trying other directions until it could move forward again. The insectoids displayed impressively complicated, insect-like behavior when set loose in a laboratory. Brooks and his insectoid robots (as well as a lion trainer, an expert on naked mole rats, and a topiary artist) were featured in the Errol Morris documentary *Fast, Cheap and Out of Control*. The film's title was borrowed from one of Brooks's best-known papers.

In the mid-1990s a group led by Brooks began building a humanoid robot that could provide an even more rigorous test of his ideas. Called Cog, it consists of a head, complete with eyes and ears, neck, arms, and torso; it looks like the robotic skeleton revealed after Arnold Schwarzenegger's flesh burned away in *The Terminator*. Cog's silicon brain has a few basic reflexes, or instincts, programmed into it, plus some learning capacity. Cog can smoothly track moving objects with its eyes alone or by moving its head too. It can "see" objects and then reach out and touch or grasp them. Like infants, it has a withdrawal reflex, which makes its arm pull away when touched.

Cog's "ears" allow it to determine the direction from which sound comes; the head then turns toward the sound. "If two people are talking, it looks back and forth between the two people," Brooks said. "It looks like it's understanding." Cog can also detect human faces and turn toward them even when they are not emitting noises. Cog has already led to some interesting findings, Brooks said. He showed me a tape of Cynthia Fell, one of Cog's designers, interacting with the robot. The tape showed Cog and Fell taking turns picking up and dropping an eraser.

Although Cog appeared to be mimicking Fell, Brooks noted that Cog did not have a mimicking program. "*She's* the one who's doing all the turn-taking here." Fell just happened to evoke a certain behavior from Cog and then turned it into a game. "That's what mothers do with babies," Brooks said. A mother turns a simple reaction of a baby into a game that then stimulates the baby into more complex behavior and learning. The lesson, Brooks suggested, is that the environment can stimulate and encourage learning, thus reducing the need for intelligence to be front-loaded into machines.

Not everyone finds Cog impressive. "It's sexier if you build a robot," Brooks's MIT colleague Steven Pinker told one reporter, "but it's not clear it's science. AI should be more disciplined, more issue-oriented, more patient." Thomas Bever, a cognitive scientist at the University of Rochester, complained, "There's so little known about the early stages of cognition that it's kind of silly to spend hundreds of thousands a year to simulate what we don't know. It's a waste of time."

Brooks himself seemed to harbor no illusions about what would come of his research. "The things we build just don't work anywhere *nearly* as well as biological systems," he lamented. He doubted whether computer programs could evolve on their own and create truly intelligent versions of themselves. That is the hope of designers of neural networks, genetic algorithms, and other alternatives to the old rule-based approach to AI. Brooks noted that all learning programs eventually hit a wall beyond which they cannot go. "There may be a theoretical maximum fitness that you can never quite get to," he said.

In fact, Brooks had come to suspect that he and other scientists trying to understand biological systems and to mimic their properties were overlooking some vital component. "My deep-seated wish," he explained, is that "we're missing something. We're not seeing something that's there. It's an elixir of life." Actually Brooks had in mind not so much a new force, particle, or vital essence but an organizational principle, concept, or language that could revitalize mind-science in the next century, just as the concept of computation had in this century. "If we had that language, we would be able to describe all these biological processes in a slightly different way, and it would give us a clue how to build imitations of those which are much more lifelike."

As usual, Brooks spiced this revelation with self-derision. He recalled that when he first mentioned his "elixir of life" notion at a workshop in Switzerland in 1995, a twenty-two-year-old graduate student from Oxford replied, "That was very interesting, what you said. I think that sentiment is quite common among scientists in the twilight of their career." At the time, Brooks was forty-one years old.

AI and Analysis

Could psychoanalysis represent the solution that Brooks and other AIers are so desperately seeking? This possibility has been broached by none other than Marvin Minsky, one of the legends of artificial intelligence. Minsky was a cofounder of MIT's Artificial Intelligence Laboratory, and like its current director Minsky is a fierce critic of the logical, rule-based approach to AI. Rule-based systems have great difficulty dealing with exceptions, according to Minsky. Minsky was fond of pointing out that the definition of a bird as a feathered animal that flies does not apply if the bird is an ostrich or a penguin or is dead or caged or has clipped wings or has its feet encased in cement or has undergone some traumatic experience that makes it psychologically incapable of flight. These arguments echoed those made by Hubert Dreyfus, the arch-enemy of AI.

Minsky has cast doubt on virtually every other approach to AI as well. Although he built one of the first neural networks in the 1950s, he later became one of the technology's harshest critics. He has also disparaged various highly mathematical "metatheories" proposed as solutions to AI. Two that were popular in the 1950s, when the field was just beginning, were cybernetics and informa-tion theory. These were followed by catastrophe theory, fractals, chaos, and complexity. "These things produce waves of enthusi-asm. They work under certain conditions," Minsky once told me. But to understand how the brain really works, he said, "you have to get beyond the metatheories."

The key to the mind's success, according to Minsky, is that it employs many different strategies for solving problems. "We have many layers of networks of learning machines, each of which has evolved to correct bugs or to adapt the other agencies to the prob-lems of thinking." Minsky knew of only one theorist other than himself who had truly appreciated this aspect of the mind. "Freud had the best theories so far, next to mine, of what it takes to make a mind," Minsky said.

Freud "recognized that the mind has many parts," Minsky elab-orated. "There are several basic instincts, maybe many." The most primitive instincts, or "machines," as Minsky referred to them, were dedicated to needs such as food, shelter, and escaping preda-tors. The superego suppresses instinctual impulses that it consid-ers inappropriate. "So Freud saw the mind as a sort of sandwich," Minsky explained. "There's this set of original goals you get from genetics, and there is a set of critical goals you get from your culture and your parents in some special way that's not known. And in between you have all the thinking."

Minsky insisted that Freud, contrary to what some modern critics asserted, had been a first-rate scientist. "Maybe you wouldn't call him a scientist, but I would," he said. Freud had realized early on that the approaches of proto-behaviorists such as Ivan Pavlov, who tried to reduce the mind to a set of simple rules, failed to do justice to the mind's complexity. Minsky dismissed the complaint that Freud did not test his theories as much as he should have. "The way you do experiments is to stick electrodes" into the brain, Minsky said. But such equipment did not exist in Freud's era. "He would have been the first to do the right experiments."

Minsky also shared Freud's belief in the therapeutic value of introspection. At the time of our interview, Minsky was working on a book called *The Emotion Machine*. "That's a person," Minsky said of the title. The book would present Minsky's ideas about "commonsense thinking and what kinds of processes it might involve." Minsky hoped the book would help people understand and thereby gain more control over their thoughts and actions. "If people have ideas about how they work," he said, "they could change themselves." Unlike Freud, however, Minsky hoped that our self-knowledge will eventually give us the power to abandon our flawed, flesh-and-blood selves and evolve into vastly superior machines. "I think the important thing for us is to grow," he said, "not to remain in our own present stupid state."

At least one young AI researcher shares Minsky's affinity for Freudian theory. Stephane Zrehen, a French-born scientist at the California Institute of Technology, believes psychoanalytic notions can be explored and tested in artificial minds. Zrehen elaborated on this notion in 1998 at a meeting of the American Association for Artificial Intelligence. In a paper titled "Psychoanalytic Concepts for the Control of Emotions in Robots," Zrehen proposed building a robotic dog. The dog's "mind" would consist of various drives—the need to eat and to defecate and the desire for affection—and an Ego, which Zrehen defined as "a mind agency in charge of finding compromises between internal drives and reality's exigencies."

Zrehen argued that "complex capacities like those attributed to the Ego can be modeled by very simple neural networks." The Ego can help the dog learn various important lessons: "After my meal I get taken out." "After my master comes home and takes off his hat and coat, he feeds me." "No matter how loud I bark, I never get fed in the morning." "When I climb on the couch, I get shouted at." "When I empty my bowels in the living room, I get shouted at." "If I empty my bowels in the street, I get patted on the head." "Future work," Zrehen concluded, "should add other key psychoanalytical concepts to the existing model of the Ego, in order to model all the elements necessary to the development of an artificial creature endowed with a psychology."

The Meaning of the Turing Test

The conflation of AI and psychoanalysis is not as farfetched as it sounds, according to sociologist Sherry Turkle of MIT. Trained as a psychoanalyst, Turkle has become a kind of shrink for the cyber-generation. In a 1988 paper titled "Artificial Intelligence and Psychoanalysis: A New Alliance," Turkle pointed out various affinities between what seem superficially to be radically different approaches to the mind. Artificial intelligence and psychoanalysis shared a common enemy, behaviorism, which treated the brain as a black box whose inner workings could be known only by studying its input and output. Both disciplines subvert conventional notions of free will and the self, psychoanalysis by emphasizing the role of unconscious processes and artificial intelligence by reducing cognition to computation. The Freudian unconscious "constitutes a decentered self," Turkle elaborated. "Inherent in AI is an even more threatening challenge: If mind is program, where is the self? It puts into question not only whether the self is free, but whether there is one at all."

In her 1995 book, *Life on the Screen*, Turkle offered a slightly different perspective on the question of whether a machine can possess a self. She recalled hearing intriguing reports about Cog, the humanoid robot being built by her MIT colleague Rodney Brooks. "Cog is controversial: for some a noble experiment that takes seriously the notion of embodied, emergent intelligence, for others a grandiose fantasy," Turkle said. She decided to visit Cog and see what the fuss was all about. She described her encounter as follows:

> Trained to track the largest moving object in its field (because this will usually be a human being), Cog "noticed" me soon after I entered the room. Its head turned to follow me and I was embarrassed to note that this made me happy. I found myself competing with another visitor for its attention. At one point, I felt sure that Cog's eyes had "caught" my own. My visit left me shaken—not by anything Cog was able to accomplish but by my own reaction to "him." For years, whenever I had heard Rodney Brooks speak about his robotic "creatures," I had always been careful to mentally put quotation marks around the word. But now, with Cog, I had found that the quotation marks disappeared. Despite myself and despite my continuing skepticism about this research project, I had behaved as though in the presence of another human being.

Cog is hardly the only product of AI to elicit this reaction from a human. An expert system developed in the 1960s named ELIZA was notorious for achieving the same results. ELIZA responded to messages typed into a computer by a human "patient" with comments that mimicked those of a psychotherapist. Its neatest trick was to extract phrases from the patient's messages and insert them into one of its stock responses. For example, if the person typed in, "I'm feeling a little anxious these days," ELIZA would respond, "Why do you think you're feeling a little anxious these days?" Any mention of such key words as *mother* would trigger the immediate response, "Tell me more about your mother" (or "father" or "dreams" or "family" or whatever else the key word was). If the program failed to recognize a pattern that enabled it to construct such a specific response, it would fall back on generic remarks: "Tell me more about that," "Please go on," or "That's very interesting."

Not only AI researchers but also some psychiatrists and pycho-analysts claimed that ELIZA's descendants would be able to supplant human therapists. The claim did not seem so grandiose, given ELIZA's ability to cast a spell over even highly sophisticated interlocutors. In *Machines Who Think*, Pamela McCorduck recalled observing an encounter at Stanford University between ELIZA and a Russian computer scientist. The exchange began simply. ELIZA asked the Russian scientist, "What brought you here to see me today?" The Russian replied, "Oh, nothing much. I'm feeling a bit tired, that's all." "Why do you think you're feeling a bit tired?" ELIZA replied.

McCorduck described what ensued: "Our visitor, an internationally respected computer scientist, who knew very well that there was a computer at the other end of the line and not a human being, nevertheless began to disclose his worries about his wife, his children, his distance—both geographical and emotional— from them. We watched in painful embarrassment, trying hard not to look, yet mesmerized all the same."

The deepest question posed by AI is whether machines can think and feel—or possess a self, as Turkle put it—as we do. In 1950 the British mathematician Alan Turing, a pioneer of computer science, proposed a simple empirical means of settling the question. On one side of a screen is a judge; on the other side are a computer and a human. The judge submits questions via teletype to the computer and the human. If the judge cannot tell which responses come from the computer and which from the human, then the computer is thinking by definition. According to the strong AI program, a machine that passes the Turing test must be conscious. But the reaction of the homesick Russian scientist to ELIZA and of the sociologist Sherry Turkle to Cog shows the flaw in the Turing test. If and when a machine convinces us that it is truly sentient, that event may say much less about the machine than it does about us. Even the most sophisticated human observers automatically attribute complex psychological states to what they know are nonsentient phenomena.

Evolutionary psychology may provide some insight into this phenomenon. One of the most intriguing notions to emerge from the field holds that all normal humans have an innate "theory-of-mind" module, which allows us to intuit the state of mind of others and thus to predict their actions. Damage to the theory-of-mind module may cause autism; autistics often seem to make no fundamental distinction between humans and inanimate objects, such as tables and chairs. But many of us have the opposite problem—an overactive theory-of-mind module. We thus unthinkingly discern sentience and complex psychological states not only within other humans and even nonhuman animals but also within rainstorms, droughts, and shooting stars. We interpret even the most random-seeming acts of nature as the acts of a wrathful or loving God. The theory-of-mind module gave us religion, and now it has given us the strong AI program.

The debate over whether machines can think and feel is just one component of the larger debate over consciousness. When the philosopher Joseph Levine coined the term *explanatory gap*, he was referring to this most mysterious of all the brain's by-products. Mind-scientists and philosophers cannot even agree on what consciousness is, let alone how it should be explained. My favorite definition of consciousness was offered by the British psychologist Stuart Sutherland in *The International Dictionary of Psychology*:

The having of perceptions, thoughts, and feelings; awareness. The term is impossible to define except in terms that are unintelligible without a grasp of what consciousness means. Many fall into the trap of equating consciousness with self-consciousness—to be conscious it is only necessary to be aware of the external world. Consciousness is a fascinating but elusive phenomenon; it is impossible to specify what it is, what it does, or why it evolved. Nothing worth reading has been written about it.

8

THE CONSCIOUSNESS CONUNDRUM

. . . .

Suppose that there be a machine, the structure of which produces thinking, feeling, and perceiving; imagine this machine enlarged but preserving the same proportions, so that you could enter it as if it were a mill. This being supposed, you might visit it inside; but what would you observe there? Nothing but parts which push and move each other, and never anything that could explain perception.

—GOTTFRIED WILHELM LEIBNIZ,
  SEVENTEENTH CENTURY

Matter can differ from matter only in form, bulk, density, motion and direction of motion: to which of these, however varied or combined, can consciousness be annexed? To be round or square, to be solid or fluid, to be great or little, to be moved slowly or swiftly one way or another, are modes of material existence, all equally alien from the nature of cognition.

—SAMUEL JOHNSON, EIGHTEENTH CENTURY

How it is that anything so remarkable as a state of consciousness comes about as a result of irritating nervous tissue, is just as unaccountable as the appearance of the Djin, when Aladdin rubbed his lamp.

—THOMAS HUXLEY, NINETEENTH CENTURY.

Searching for a "molecular" explanation of consciousness is a waste of time, since the physiological processes responsible for this wholly private experience will be seen to degenerate into seemingly quite ordinary, workaday reactions, no more and no less fascinating than those that occur in, say, the liver.

—GUNTHER STENT, TWENTIETH CENTURY

Over the course of his or her career, every science reporter receives letters from people who claim to have discovered something extremely important: a theory that proves Einstein's view of time and space was wrong, a reformulation of quantum mechanics, a one-page proof of Fermat's last theorem, or simply the key to everything. The writers are generally unaffiliated with any formal institution (prisons and mental hospitals don't count) and desperate for recognition. Some missives are obviously products of clinical derangement. The manuscript may be dozens or even hundreds of pages long, hand-written, packed with esoteric words and mathematical notation, uppercase letters, and exclamation marks. Often the color of the pen or pencil keeps changing. God (or, even worse, Wittgenstein) keeps rearing his head. The most disturbing letters are those that are most lucid and intelligent and erudite. The authors use conventional typography, cite legitimate sources, employ formulas judiciously, construct their arguments carefully, minimize references to God or Wittgenstein. But something is still a little . . . off.

I have learned through experience that it is best not to respond to these letters. Amused by one that presented—ironically, I thought—a method for faster-than-light transportation (the letterhead identified the writer as president of Transluminal Industries, Inc.), I wrote the author to ask him—ironically, I thought—if I could buy stock in his company. The letters, faxes, and phone calls lasted for weeks before subsiding. But I still feel uneasy simply ignoring these would-be Newtons and Darwins and Einsteins. It is not that I think one of them might actually have found the key to everything. Rather, the thought of these lonely truth seekers seeking validation, or at least a response, from the indifferent world stirs even my hardened heart. I have sometimes wished I could put each person who has discovered the key to everything in touch with all the others who have found a different key. Ideally, they could have a conference, exchange views, and decide whose key worked best—although of course this resolution would be extremely unlikely.

Such a conference, I suspect, would resemble one that I attended in April 1994. "Toward a Scientific Basis of Consciousness" was billed (erroneously, as it turned out) as the "first American interdisciplinary scientific conference on consciousness." Held at the University of Arizona in Tucson, the meeting was a kind of distillation, or caricature, of the entire enterprise of mind-science, with all its fractiousness and confusion amplified a thousandfold. Virtually every scientific discipline (and not a few pseudoscientific or even antiscientific ones) was represented: psychology, psychiatry, neurology, neuroscience, artificial intelligence, mathematics, chaos theory, physics, and of course philosophy. Many of the attendees were eminent professors at major institutions, such as the University of Oxford and the California Institute of Technology. Their presentations had all the trappings of serious scientific discourse: technical terms, references to experimental data, equations. But they still seemed a little . . . off.

Not that the meeting was dull. The tone was established early on by its chief organizer, Stuart Hameroff, an anesthesiologist at the University of Arizona and a member of the quantum-consciousness gang. Looking out over the motley crowd during his welcoming speech, Hameroff, a middle-aged hipster sporting a goatee and ponytail, exclaimed, "Wow, this is just like Wood-stock!" Later he presented the best slide of the conference. It showed a man with long stringy hair and bulldog features staring straight into the camera, his eyes bulging and his teeth clenched in rage and pain. A long piece of steel protruded from both temples. Hameroff explained that the man was a prisoner who had been stabbed in a drug-related dispute. Far from killing this modern-day Phineas Gage, the steel bar had not even rendered him unconscious. He was knocked out only after physicians administered an anesthetic at a local hospital. The incident, Hameroff said, showed how robust consciousness can be. Another lesson, he added, was that "the Pentothal is mightier than the sword."

Every conceivable key to consciousness had its adherent in Tucson. Steen Rasmussen, a Danish physicist associated with the Santa Fe Institute, headquarters of the trendy field of complexity, suggested that consciousness might be an "emergent"—that is, unpredictable, irreducible, and holistic—property of the brain's complex behavior, just as superconductivity is an emergent property of certain ceramic compounds at relatively high temperatures. He proposed that the concept of free will might be explained by "downward causation." Scientific explanations, Rasmussen elaborated, often imply that causation only happens "upward"; that is, the large-scale behavior of a system is determined by the behavior of its smallest parts. But causation can also happen downward; an emergent phenomenon like a mind is to some extent independent of the small-scale processes that created it and can even exert top-down control over them. Hence, free will.

The veteran neuroscientist Karl Pribram, who spent thirty years at Stanford before moving to Radford University in 1989, updated a once-popular theory that he developed in the 1960s. Pribram had speculated that memory might work in a manner similar to holography, in which a three-dimensional image is generated by the interference of two overlapping laser beams. One of the most striking features of a hologram is that each part of the image incorporates the whole image, albeit at a lower resolution; one can thus reconstruct the entire image from a minute component.

Pribram pointed out that in the same way, minute mnemonic cues can yield entire memories.

A weakness of Pribram's original holographic model had been that there appeared to be no neural equivalent to the laser beams that generate holograms. But since then researchers have found that large groups of neurons often fire repeatedly in step and at the same frequency, like the waves of light in a laser. Pribram suggested that these oscillating neurons generate minute electrical fields that serve as analogues of light waves in a hologram; interference and resonance between the overlapping fields may help to generate memory, perception, and other mental properties.

Danah Zohar, who earned a degree in physics from MIT and then studied philosophy and religion under the psychoanalyst Erik Erikson at Harvard, reiterated the main theme of her 1990 book, *The Quantum Self*. It was time to move beyond dualism, she said, and accept that both matter and mind stem from a deeper source, "the quantum." Human thoughts, she assured us, are quantum fluctuations of the vacuum energy of the universe, which "is really God." After her talk, an audience member noted that physicists were discovering deep links between quantum mechanics, information theory, thermodynamics, and black holes. These findings, he said, might also yield clues to the mystery of consciousness. "There's not a black hole in our brains," he added, "but . . ." "I think there *is* a black hole in my brain!" a presumably overloaded listener interrupted.

The conference was not devoid of empirical findings, but few yielded unambiguous meanings. Benjamin Libet, a psychologist at the University of California at San Francisco, described an experiment in which subjects were asked to flex a finger at a moment of their choosing while noting the instant of their decision as displayed on a clock. Data from sensors on the fingers showed that the subjects took 0.2 second on average to flex their fingers after they had decided to do so. But an electroencephalograph monitoring their brain waves revealed that the subjects' brains typically generated a spike of activity 0.3 second *before* they consciously made the decision to push the button.

"The actual initiation of volition may have begun even earlier in a part of the brain we weren't monitoring," Libet commented. A physician from California (who earlier had been handing out copies of a self-published book that provided advice on "how to be happy") asked Libet whether his findings related to the question of free will. "I've always been able to avoid that question," Libet replied with a grimace. He cautiously suggested one implication of his work: we may exert free will not by initiating intentions but by vetoing, acceding, or otherwise responding to them after they well up from our unconscious.

Other presentations concerned people whose brains—and thus minds—had been damaged through disease or trauma. One worker showed how Alzheimer's patients in the early stages of disease pass through a period during which they become excruciatingly aware of their memory loss. As their memories continue to deteriorate, they become less cognizant of their loss. Several researchers discussed brain-damaged patients afflicted with a strange syndrome called blindsight. Such patients are subjectively blind, but at some level their brains are still receiving and processing visual information. If you place a picture of, say, a lion in front of a man suffering from blindsight, he will assert that he cannot see anything. However, if you insist that he guess what the picture shows, he will often guess correctly. Blindsight suggests that perception and consciousness are to some extent separate phenomena underpinned by different neural regions.

A neurologist from the University of North Dakota showed a video of a young woman suffering from epilepsy so severe that surgeons had severed the neural cables linking the two hemispheres of her brain. Although the surgery had alleviated her epilepsy, it left her with two centers of consciousness vying for dominance. When asked if she had any sensation in her left hand, which was neurally connected to only one hemisphere, she shouted, "Yes! Wait! No! Yes! No, no! Wait, yes," her expression agonized. The researcher then handed her a sheet of paper with the words *yes* and *no* written on it and asked her to point to the correct answer. The woman stared at the sheet for a moment. Then her left forefinger stabbed at "Yes" and her right forefinger at "No." A psychotherapist in the audience suggested afterward that even healthy people experience some fracturing of the self. Another proposed that the woman's two selves might be trained to get along better through conflict resolution.

Those who found the formal lectures too bland could forage in the hallway outside the auditorium for even more farfetched fare. "That's where the real action is," said a journalist who sported a nose ring and six-inch chin braid and was covering the meeting for an obscure Internet magazine. At one point I found myself engaged in a hallway discussion with a large, bald man wearing what appeared to be linen pajamas. He was disappointed, even angry, that almost everybody at the conference defined consciousness in such narrow terms; consciousness was obviously a property not just of humans and other higher animals but of all natural phenomena, including bugs, plants, and rocks. I suggested that scientists would not get very far in understanding consciousness if they defined it in such a flexible manner. The pajama-clad man heatedly told me that he, like me, used to be stuck in an extremely narrow, materialistic paradigm. He had grown out of it, and if I wasn't so close-minded I might too.

Christof Koch's Consciousness

If one had to seed the various contenders within this scientific free-for-all, the top position would probably go to Christof Koch, a German-born neuroscientist from the California Institute of Technology. Koch should arguably receive much of the credit—or blame—for the burgeoning scientific interest in consciousness. In 1990 he and Francis Crick, co-discoverer of the double helix and one of this century's most formidable scientists, announced in a jointly written paper that it was time to make consciousness a subject of serious scientific inquiry. Contrary to the assumptions of psychologists, philosophers, and others, Crick and Koch asserted, one cannot hope to achieve true understanding of consciousness or any other mental phenomena by treating the brain as a black box. Only by examining neurons and the interactions between them could scientists create models that are truly scientific, like those that explain heredity in terms of DNA. Crick elaborated on these ideas in his 1994 book *The Astonishing Hypothesis.* Crick dedicated the book to Koch, "without whose energy and enthusiasm this book would never have been written."

Koch's energy and enthusiasm were evident in his lecture at Tucson. A tall, lean man with a staccato, German-accented patter, he loped around the stage as he spoke, pausing only to flash a slide or crack a joke. He reminded the audience that he and Crick had defined consciousness as the ability of the brain to focus on, or attend to, one set of phenomena out of all those that are impinging on the mind. The question then becomes a seemingly simple one: How does the brain attend to, or become aware of, say, a single face in a roomful of people?

The problem is actually quite complicated, Koch explained, because even a single visual scene is processed by many different parts of the brain. "There is no single place where everything comes together," Koch said. He added that the apparently distributed nature of awareness makes evolutionary sense, since the brain is less susceptible to being shut down by a single, localized blow. But what mechanism transforms the firing of neurons in numerous regions of the brain into a unified perception? "This is the binding problem," Koch explained. By solving the binding problem (what I call the Humpty Dumpty dilemma), he suggested, neuroscientists might go far toward solving the problem of consciousness.

One possible solution to the binding problem had been suggested by experiments showing that neurons in different parts of the brain occasionally fire at the same frequency—roughly forty times per second. Koch asked the audience to imagine the brain as a Christmas tree with billions of lights flickering apparently at random. These flickerings represent the response of our visual cortex to a roomful of people. Suddenly a subset of those lights flickers at the same frequency, forty times per second, as the mind attends to, or becomes conscious of, a single face. Koch conceded that the evidence for 40-hertz oscillations is tenuous; it had shown up most clearly in anesthetized—that is, unconscious—cats. Another form of binding could be simple synchrony: neurons merely fire at the same time and not necessarily the same frequency. Evidence for synchrony was also trickling in from animal experiments, according to Koch.

One onlooker who objected to Koch's presentation was Walter Freeman, a tall, thin, white-bearded neuroscientist at the University of California at Berkeley. (Freeman's father was the neurosurgeon who popularized lobotomies in the United States in the 1950s.) Freeman's criticism was significant because he was one of the first investigators of 40-hertz neural oscillations. Although these oscillations may play some role in consciousness, Freeman said, they are unlikely to be the key to the problem, any more than oxygen uptake or blood flow or other ubiquitous phenomena. Forty-hertz oscillations are "a blind alley, a dead end," Freeman said. "The current wave of enthusiasm is misplaced."

Freeman himself advocated a more complex model of awareness based on chaos theory. Chaotic systems seem random but actually conceal a hidden order, which can be described with mathematical objects known as attractors. Chaotic systems display what is called sensitivity to initial conditions, or, more colorfully, the butterfly effect. In principle the fluttering of a butterfly's wing in, say, Iowa can trigger a cascade of effects that culminates in a monsoon in India. By plotting the firing patterns of large groups of neurons, Freeman has shown that they form chaotic patterns. This behavior might account for the brain's ability to respond to complex perceptual data with astonishing rapidity, Freeman hypothesized. The sight of a familiar face, for example, might trigger an almost instantaneous shift in the chaotic pattern of a group of neurons in the visual cortex, resulting in recognition. But even Freeman conceded that his theory is only—at best—one piece of the puzzle.

Owen Flanagan, a philosopher at Duke University, concurred that no single mechanism—whether Freeman's chaotic neural behavior or Koch's 40-hertz oscillations—is likely to solve the consciousness conundrum. There may be as many modes of consciousness, Flanagan contended, as there are modes of memory and perception. Our awareness of a cat's odor may stem from an entirely different set of neurons and neural processes than our visual glimpse of the same cat. Flanagan argued on behalf of a philosophy called constructive naturalism, which holds consciousness to be a common biological phenomenon occurring not only in humans but in many other animals, and almost certainly in the higher primates. A full understanding of consciousness would emerge from what Flanagan called a process of triangulation, which combines reports from human subjects about their subjective experience with objective data from both psychology and neuroscience.

"Listen carefully to what individuals have to say about how things seem," Flanagan elaborated.

> Also, let the psychologists and cognitive scientists have their say. Listen carefully to their descriptions about how mental life works, and what jobs if any consciousness has in its overall economy. Third, listen carefully to what the neuroscientists say about how conscious mental events of different sorts are realized, and examine the fit between their stories and the phenomenological and psychological stories. Now this triangulation procedure will, I claim, yield success in understanding consciousness if anything will.

Roger Penrose's Quantum Leap

Flanagan's approach was far too conventional for many speakers in Tucson, who favored more radical models involving quantum mechanics. Physicists and philosophers began speculating that consciousness might be linked in some profound way to quantum mechanics soon after its inception early in this century. According to certain interpretations of Heisenberg's uncertainty principle, the act of measurement—which ultimately involves a conscious observer—has an effect on the outcome of quantum events; electrons act like waves in one experiment and particles in another. Quantum theory implies that particles can pass, ghost-like, through walls and be in two places at the same time. Quantum mechanics also allows so-called nonlocal effects, in which two particles exert subtle influences on each other at speeds faster than that of light. Einstein, who never fully accepted the existence of nonlocal effects, once derided them as "spooky action at a distance."

Quantum theories of consciousness involved little more than hand-waving, however, before Roger Penrose began considering them in the late 1980s. Penrose, a mathematical physicist at the University of Oxford, is a star of twentieth-century science. He first made his reputation as an authority on black holes and other gravitational phenomena. In the 1970s he and his student Stephen Hawking (who later went on to write the best-seller *A Brief History of Time*) constructed a mathematical proof that all black holes contain a singularity—a point at which the density of matter approaches infinity and the laws of conventional physics break down.

In 1989, Penrose declared in *The Emperor's New Mind* that virtually all mainstream approaches to understanding the mind—including those emerging from artificial intelligence, cognitive science, and neuroscience—are wrong-headed. The key to Penrose's argument was the incompleteness theorem set forth by the mathematician Kurt Gödel in the 1930s. Gödel proved that any system of axioms complex enough to generate arithmetic is incomplete; that is, the system will yield "undecidable" propositions, whose truth or falsity cannot be established with those axioms alone. These undecidable propositions are often mathematical versions of such well-known, self-referential paradoxes as the statement, "I am lying." One can resolve undecidable propositions by adding new axioms to the system, but these new axioms will generate a new set of undecidable propositions, ad infinitum.

The meaning of Gödel's theorem, Penrose suggested, is that

mathematics can never be reduced to an algorithm or set of rules that churns out theorems and proofs. Penrose offered his own subjective experience as a mathematician as evidence for this proposition; his best work arose not from any deductive, logical process but from sudden intuitions and insights into an indescribably beautiful Platonic realm. Penrose concluded that no mechanical, ruled-based system—that is, neither classical physics, computer science, nor neuroscience as currently construed—can account for the mind's creative capacity. Human cognition must derive from some more subtle effects, probably related to quantum mechanics, that so far had eluded the gaze of conventional science.

In his talk at Tucson, Penrose summarized the major arguments of a new book, *Shadows of the Mind*, that followed up on the ideas he had set forth in *The Emperor's New Mind*. He prefaced his remarks by reporting that Deep Thought, a computer that had beaten some of the world's greatest chess players, was still stumped by problems that even an amateur player could solve. "What computers can't do is understand," Penrose asserted. (Deep Thought was a predecessor of the IBM computer Deep Blue, which went on to defeat the world chess champion Gary Kasparov in 1997.)

Penrose declared that the talents of the human mind could be explained only by a still-undiscovered physical theory that incorporates both quantum mechanics and relativity theory. The effort to find such a theory—which is referred to as a quantum-gravity theory, a unified theory, or a theory of everything—is the ultimate goal of theoretical physics. In *The Emperor's New Mind*, Penrose had avoided specifying where and how these quasi-quantum effects might work. He now suggested that quantum nonlocality (the ability of one part of a quantum system to affect other parts instantaneously) might be the solution to the binding problem. These spooky quantum effects might come into play at the level of microtubules, minute tunnels of protein that serve as a kind of skeleton for cells.

Penrose's talk delighted Stuart Hameroff, the anesthesiologist who organized the Tucson meeting. It was Hameroff who first proposed in the 1980s that microtubules might be the seat of consciousness; his papers led Penrose to endorse the hypothesis. Hameroff claimed to have found evidence that anesthesia arrests consciousness by hindering the motion of electrons in microtubules. Microtubules, he concluded, generate consciousness by carrying out nondeterministic, quantum-based computations. Each neuron is thus not merely a switch but a powerful computer. "Most people think of the brain as 40 billion switches," Hameroff once told me, referring to the total number of neurons in the brain. "But we think of it as 40 billion tiny computers."

Hameroff managed to cram his talk on quantum consciousness with every conceivable scientific buzzword: *emergent, fractal, self-organizing, dynamical.* Other quantum-consciousness participants made Hameroff look like a paragon of rigor. A group of British researchers claimed that they had found tentative evidence that cognition does indeed exploit quantum effects. The group had tested the ability of subjects to perform simple tests while their brain waves were monitored with an electroencephalograph (EEG). According to the investigators, the performance of certain subjects varied depending on whether the EEG was turned on or off. Their conclusion? When the machine was turned on, it "observed" the brain and therefore changed the course of its thoughts, just as observing an electron passing through an interferometer alters its properties.

Conscious Thermostats

Although proponents of quantum consciousness abounded in Tucson, so did critics. Christof Koch summed up the quantum-consciousness thesis in a syllogism: Quantum mechanics is mysterious, and consciousness is mysterious; therefore they must be related. John Taylor, a physicist and neural network researcher at King's College London, insisted that Penrose and other quantum-consciousness enthusiasts were ignoring the most basic facts about quantum mechanics. For example, nonlocality and the other exotic quantum effects that are supposedly crucial to consciousness are generally observed only at temperatures near absolute zero—or at any rate far below the ambient temperature of most living brains.

Taylor also objected to the quantum hypotheses on pragmatic grounds. Nuclear physics had had absolutely no relevance to biology thus far. Before turning to the extremely reductionist, *sub*nuclear approach Penrose and others advocate, researchers should explore possibilities that are more plausible and experimentally accessible—and that have already been modestly successful at explaining certain features of memory and perception. "If that fails, then maybe we should look elsewhere," Taylor said.

Both quantum-consciousness theories and neural ones were rejected in Tucson by David Chalmers, a young Australian philosopher and mathematician. In his lecture, Chalmers declared that physical theories can account only for the various *functions* of the brain, such as perception, memory, and decision-making. But no physical theory can explain why these cognitive functions are accompanied by conscious sensations, which some philosophers call *qualia*. Chalmers called consciousness "the hard problem."

Chalmers believed he had found a potential philosophical solution to the hard problem, however. Just as physics assumes the existence of fundamental properties of nature such as space, time, energy, and mass, so a theory of consciousness must posit the existence of a new fundamental property: information. Information, Chalmers elaborated, always has some physical embodiment, such as the arrangement of ink spots on a piece of paper or of electrons in a computer. But information is not *purely* physical; it also has a "phenomenal" aspect. (*Phenomenal* is a philosopher's term that is roughly equivalent to *subjective* or *experiential*.)

According to his theory, Chalmers noted, any object that processes information must have some conscious experience. "Where there is simple information processing, there is simple experience, and where there is complex information-processing, there is complex experience. A mouse has a simpler information-processing structure than a human, and has correspondingly simpler experience; perhaps a thermostat, a maximally simple information-processing structure, might have a maximally simple experience?" In later writings, Chalmers boldly answered this question in the affirmative. If one accepts his information-based hypothesis, one must accept that a thermostat is indeed conscious. In effect, Chalmers was espousing the same philosophy as the bald-headed man in the white pajamas who had harangued me in the hallway: almost *everything* in the universe might be in some sense conscious.

Consciousness Explained Away

Although all speculations about consciousness seem a bit off, some are more off than others. One source of confusion in the debate over consciousness is that different people define consciousness in different ways. To those of a New Age bent, such as Danah Zohar, consciousness means self-consciousness, or even a kind of mystical hyperawareness. But these capacities are manifest only in a few rare humans, and only occasionally even in them; moreover, self-consciousness is just a special type of consciousness, in which the object of awareness is the self.

When Roger Penrose talks about consciousness, he is generally referring to the ability to solve extremely complex problems, especially those of a mathematical nature. Most people would call this trait intelligence, and intelligence of a very rare kind. Penrose's books make the case that computers as currently construed cannot replicate this type of highly logical intelligence. Ironically, computers are much more capable of logical, precise reasoning ability—the basis not only of mathematics but also of chess and other games—than ordinary common sense (as I tried to show in the previous chapter). After all, artificial intelligence researchers have had some modest success with programs that can pose and prove theorems, and a computer did trounce the world chess champion in 1997.

The most sensible definition of consciousness is the one favored by Christof Koch and other researchers: consciousness is simply sentience, or awareness, and it is a phenomenon that probably occurs not only in humans but almost certainly in many higher animals as well. Consciousness is also a by-product of specific physical processes occurring in specific types of matter; without this special type of matter, consciousness cannot exist. This view of consciousness leads inevitably to a rather brutal materialism, one that rejects any scheme placing mind and matter on an equal footing or giving primacy to mind. To put it bluntly, we have all seen bodies without minds, but only mystics, psychics, and psychotics have seen minds without bodies.

To the best of our knowledge, the universe existed for billions of years before life emerged on our little planet. Billions more years passed before algae and other single-celled organisms on earth evolved into multicellular organisms, such as slime molds and *Tyrannosaurus rex*. And only in the last instant of geological time has life been conscious enough to ponder consciousness. So far science has been unable to find extraterrestrial life. If an asteroid obliterates all life on earth tomorrow, the entire universe may well become devoid of life and, hence, consciousness. But the universe will get along quite well without us. It will hurtle blindly along its path until the end of material time, unless or until it gives birth to sentient life once again. To believe otherwise is, well, narcissistic.

The neural approach to consciousness advocated by Koch, Crick, and others may one day be validated by experiments on both humans and animals. By studying blindsight, anesthesia, and other phenomena, researchers may isolate the neural events that are both necessary and sufficient for *human* consciousness. This knowledge might have practical consequences. It might provide insights into schizophrenia and other cognitive disorders; it might yield more effective anesthetics and analgesics; it might even show artificial intelligence researchers how to make their machines more like us.

The key to consciousness might be some relatively simple

neural mechanism, such as the 40-hertz oscillations that Koch has very tentatively proposed. But even Koch concedes that the answer will probably turn out to be much more complicated, and hence much less satisfying, than he hopes. As the philosopher Owen Flanagan has emphasized, there are many different types of memory, emotion, perception, and intelligence, and there may also be many types of human consciousness, each arising from different neural processes. In fact, as the Harvard psychologist Howard Gardner has argued, we might understand consciousness only in an intuitive, literary sense, which according to many scientists means that we might not really understand it at all.

A scientific explanation of human consciousness will surely not resolve all our debates over consciousness, because it will leave too many questions unanswered. One unanswerable question is this: What conditions are both necessary and sufficient for consciousness to occur not only in humans but in *any* collection of matter? Here we are bumping into one of the oldest riddles of philosophy, the solipsism problem. My dictionary defines *solipsism*, rather awkwardly, as "a theory holding that the self can know nothing but its own modifications and that the self is the only existent thing."

Solipsism is a radical skepticism inspired by the recognition that each of us is trapped within a seamless prison of subjectivity. None of us can be absolutely sure that anyone else possesses consciousness, or sentience, or an inner life. We all make this assumption because it is the reasonable thing to do—and perhaps because we are compelled to do so by our innate theory-of-mind module. Most reasonable people, and even many neuroscientists and philosophers, also assume that apes, monkeys, and other mammals with relatively large brains have conscious experience. But it is worth recalling that some very smart humans, notably Descartes, have believed that all nonhuman animals are nonsentient automatons. The point is that the solipsism problem prevents us from empirically resolving the matter one way or the other.

Disagreement intensifies as one moves to creatures less like us, such as bees and bass and barnacles. Is a cortex required for sentience, or just a brain, or will a simple nervous system like the one possessed by Eric Kandel's sea snails suffice? Must the information-processing circuitry consist of organic substances, such as proteins and nucleic acids and neurotransmitters? Or can it be made of copper and mercury—like David Chalmers's thermostat—or of silicon, like a computer chip? If or when computers become capable of chatting with us like old friends, many of us may be inclined to grant them consciousness. But reasonable people always can and will disagree, because there is no way to settle the dispute empirically. As I heard Koch blurt out to David Chalmers at a cocktail party in Tucson, "How do I even know you're conscious?"

Science is also unlikely to "explain" free will, the riddle wrapped in the enigma of consciousness. From one perspective, the existence of free will is obvious. Some creatures have a greater ability than others to perceive different options and to choose among them. Humans have more of this ability than cats and dogs. Mentally healthy humans have more than schizophrenics or obsessive compulsives. Adults have more than five-year-old children. And five-year-old children have more than infants. Surely free will must exist if some organisms have more of it than others.

Brain damage can also eliminate our sense of free will. In a postscript to *The Astonishing Hypothesis*, Francis Crick recalled reading an article about a brain-damaged woman who temporarily lost her ability to act upon her intentions or even to form intentions. After she recovered, she recalled feeling empty, sensitive to but unable to respond to any external stimulus. She understood what others were saying to her but had nothing to say in reply. Crick was "delighted" to learn that the woman's brain had been damaged in the anterior cingulate sulcus, an area that other experiments had shown "received many inputs from the higher sensory regions and was at or near the higher levels of the motor system." Crick immediately announced to several colleagues that "the seat of the Will had been discovered!"

And yet Crick himself has suggested that free will may be an illusion created by our imperfect self-knowledge. (Crick cited the experiments of Benjamin Libet, described previously in this chapter, as tentative evidence for this position.) Crick once pointed out to me that even the simplest act is the culmination of a vast amount of neural activity unfolding below the level of our awareness. "What you're aware of is a decision, but you're not aware of what makes you do the decision. It seems free to you, but it's the result of things you're not aware of."

I find this line of reasoning all too persuasive. This afternoon, I may choose to stop working earlier than usual to take my two kids for a walk in the woods. But just how free would such a choice be? Each supposedly free act is the culmination of an infinite sequence of proximate and long-range causes. Quantum mechanics and chaos theory suggest that pinpointing the causes might be extremely difficult, even impossible, but that does not mean the causes do not exist. Retracing the steps that led to a particular act takes us back into childhood and the womb, back through the history of *Homo sapiens* and of all life on earth, and finally to the big bang itself, the creation event that supposedly set everything in motion. I didn't ask for any of this, so how free can I be? If free will is an illusion, however, it is an absolutely necessary one, more so even than God. As William James once wrote, "My first act of free will shall be to believe in free will."

Mysterianism Ascendant

My position on consciousness and free will has been called *mysterianism*. The mysterian position is an old and venerable one, as the quotations at the beginning of this chapter demonstrate, but the term was coined only recently. In his 1991 book *The Science of the Mind*, the philosopher Owen Flanagan noted that some modern scientists and philosophers have suggested that consciousness might never be completely explained in conventional scientific terms—or in any terms, for that matter. Flanagan dubbed these modern doubters "the new mysterians," after the 1960s rock group Question Mark and the Mysterians. (The term did not originate with the rock band; *The Mysterians*, a low-budget Japanese film about an alien invasion, was released in 1959.)

In defending their position, mysterians often borrow a line of reasoning from Noam Chomsky. The MIT linguist has distinguished between *problems*, which seem solvable at least in principle through conventional scientific methods, and *mysteries*, which seem insoluble even in principle. Chomsky noted that all organisms have certain capacities and limitations that result from their particular biology. Thus, a rat might learn how to navigate a maze that requires it to turn right at every juncture or to alternate between right and left; but no rat will ever learn to navigate a maze that requires it to turn left at every juncture corresponding to prime numbers. That talent exceeds its cognitive capabilities. In the same way, certain problems addressed by science may lie forever beyond our capacity for understanding. These are mysteries, now and possibly forever.

Chomsky has implied in various writings that he considers consciousness, free will, and other aspects of the mind to be mysteries. Yet in a conversation with me, Chomsky once took issue with a fundamental tenet of the mysterian position. "There is no such thing as the mind-body problem," Chomsky asserted. "For there to be a mind-body problem, there has to be some characterization of body, and Newton eliminated the last conception of body anybody had." Newton, Chomsky explained, is supposedly the progenitor of the mechanistic, materialist worldview that gave rise to the mind-body problem. But Newton's own theory of gravity, which showed that objects can influence each other in nonmechanistic ways, actually *shattered* the materialist worldview.

Materialism, Chomsky elaborated, presupposes that the world consists of objects that interact through direct contact with each other. But Newton, by discovering gravity—action at a distance—showed that materialism doesn't work even for a phenomenon as simple as a ball rolling down a plane. The world consists not of material objects influencing each other through direct contact but of immaterial *properties*. These properties include gravity, electromagnetism, and, yes, consciousness. "It's an interesting element of the history of human irrationality, that people continue to talk about the mind-body problem," Chomsky said. "I should say I haven't convinced a lot of people," he added.

Of course, Chomsky has not really resolved the mystery of consciousness with this argument; rather, he has pointed out that consciousness is only one of many mysterious properties of nature. A different objection to mysterianism has been advanced by the philosopher Daniel Dennett of Tufts University. Dennett has accused mysterians of taking a position akin to vitalism, the hoary notion that life springs not from purely physical processes but from some ineffable *élan vitale*. Just as vitalism vanished after biologists discovered DNA-based replication and other fundamental biological mechanisms, Dennett argued, so will the mysterian position fade away once neuroscientists account for attention, short-term memory, and other key mental functions. Consciousness is merely the sum of these cognitive functions, just as life is the sum of replication, metabolism, and other biological processes.

But as patients with blindsight demonstrate, consciousness is to some extent a distinct cognitive phenomenon. Another flaw in Dennett's position is his implication that science has really *explained* life—that is, rendered it devoid of mystery. As I pointed out at the end of Chapter 6, life remains profoundly mysterious, even when "explained" by evolutionary biology and Mendelian genetics and molecular biology. All our scientific knowledge cannot tell us—now and perhaps forever—whether life exists elsewhere in the universe or only here on our lonely little planet. Science cannot really tell us why life appeared on earth and why it gave rise to creatures like us. We do not know whether life was a highly probable and even inevitable consequence of the laws of physics and chemistry or a once-in-eternity fluke. The mystery of *conscious* life—and particularly life conscious enough to ponder consciousness—is still more acute.

Mysterianism is becoming a mainstream position. Among those who have publicly embraced mysterianism is Steven Pinker, the MIT psycholinguist and evolutionary psychologist. At the end of *How the Mind Works*, which otherwise epitomizes scientific triumphalism, Pinker concluded that consciousness, free will, the self, and other riddles posed by the mind are probably unsolvable:

> Our minds evolved by natural selection to solve problems that were life-and-death matters to our ancestors, not to commune with correctness or to answer any question we are capable of asking. We cannot hold ten thousand words in short-term memory. We cannot see in ultraviolet light. We cannot mentally rotate an object in the fourth dimension. And perhaps we cannot solve conundrums like free will and sentience.

Even the neuroscientist Christof Koch has admitted that in the long run, the mysterians could turn out to be right. In his lecture in Tucson, Koch conceded that a neural theory of consciousness might not solve such ancient philosophical conundrums as the mind-body problem and the question of free will; these riddles might simply be beyond the scope of science. To emphasize this position, Koch flashed a quotation from the end of Ludwig Wittgenstein's oracular monograph, *Tractatus Logico-Philosophicus*: "Whereof one cannot speak, thereof one must be silent." Koch received an even bigger laugh from the audience when he paraphrased "another giant of the twentieth century," Dirty Harry, the tough cinematic cop played by the actor Clint Eastwood: "A scientist has gotta know his limitations."

The Mystical Road to Knowledge

By now it should be obvious that those debating whether consciousness can be explained invest not only the term *consciousness* with different meanings but also the term *explanation*. For hardnosed types such as Daniel Dennett, a physiological—and, more precisely, neural—model will almost certainly suffice. When *these* regions of the brain perform *these* functions, consciousness results. Others want something more than 40-hertz oscillations in

the anterior cingulate sulcus. They yearn for an insight so powerful that it will instantly dispel the mystery from consciousness,
like the sun burning off a morning fog. They seek not just an explanation but a *revelation*.

Even some hard-core mysterians, such as the philosopher Colin
McGinn of Rutgers University, do not rule out the possibility of
such a revelation. McGinn once suggested to me that if artificial
intelligence researchers succeed in creating truly intelligent machines, the machines might have insights into consciousness that
elude mere humans. "It doesn't seem to me at all impossible,"
McGinn said, "that you could engineer a device that's operated by
completely different principles and was capable of reconceptual-
izing things. And who knows what it could do."

Ironically the AI visionary Marvin Minsky has expressed
doubts about whether our cyber-descendants will solve the riddle
of their own minds. "When intelligent machines are constructed,"
Minsky wrote, "we should not be surprised to find them as confused and stubborn as men in their convictions about mind-
matter, consciousness, free will and the like." In other words, even
if AIers succeed in *replicating* the mind—creating machines that
not only mimic but improve on our cognitive powers—the mind
may still remain unexplained.

Optimists hope that even we ordinary humans might reach a
deeper understanding of our minds through such mind-altering
methods as meditation, fasting, and psychedelic drugs. One speaker
who made this point in Tucson was Andrew Weil, the alternative-
medicine guru and best-selling author. Weil told the audience
about a group of Peruvian shamans who reportedly see identical
visions upon ingesting a psychedelic substance; in effect they
share the same consciousness. Weil said that he and a friend had a
similar experience of being "in each other's consciousness" when
they smoked the dried venom of the Colorado River toad, *Bufo al-
varius*. These bizarre mystical experiences, Weil suggested, might
yield the kind of insights into consciousness needed to break the
current theoretical impasse.

Like most other positions in the debate over consciousness, the idea that mystical experience can supplement empirical investigations is an old one. In *Civilization and Its Discontents* Freud described a friend who had reported "a sensation of 'eternity,' a feeling as of something limitless, unbounded—as it were, 'oceanic.'" Freud admitted that he had never had such sensations himself, but he doubted whether they could offer any genuine insights into reality. Those who experience mystical sensations, he proposed, are reliving their infancy, when they have not yet learned to distinguish between themselves and the external world. As for the belief of many mystics that their experience has put them in touch with a higher power, Freud commented drily, "I cannot think of any need in childhood as strong as the need for a father's protection."

William James offered a more sympathetic perspective in *The Varieties of Religious Experience*. James acknowledged that mystical experiences often occur to those who are in distress or suffer from neuropathology of some sort; mystical states can also be artificially induced by drugs such as ether or nitrous oxide. But just because a mystical experience has a physiological or even pathological basis, the founder of pragmatism asserted, does not mean that its insights are invalid. According to this reasoning "none of our thoughts and feelings, not even our scientific doctrines, not even our dis-beliefs, could retain any value as revelations of the truth, for every one of them without exception flows from the state of its possessor at the time."

That was not to say, James emphasized, that all mystical revelations should be accepted as true. They should be judged according to "what we can ascertain of their experiential relation to our moral needs and to the rest of what we hold to be true." At the very least, James said, mystical experiences should force us to accept how little we really know about our own minds. James's own experiments with nitrous oxide left him convinced that

> our normal waking consciousness, rational consciousness as
> we call it, is but one special type of consciousness, whilst all

about it, parted from it by the flimsiest of screens, there lie potential forms of consciousness entirely different. We may go
through life without suspecting their existence; but apply the
requisite stimulus, and at a touch they are there in all their com-
pleteness, definite types of mentality which probably somewhere have their field of application and adaptation. No
account of the universe in its totality can be final which leaves
these other forms of consciousness quite disregarded....
[T]hey forbid a premature closing of our accounts with reality.

A major theme of James's book was that mystical experiences
can take radically different forms. He demonstrated this point by
providing scores of first-person accounts. Some narrators described the "oceanic" ecstacy that Freud had disparaged, but others were left terrified and alienated. In one of the most harrowing
accounts in James's book, the narrator was alone in his dressing
room when he abruptly remembered an epileptic patient whom
he had once seen in an insane asylum. The patient had been a
"black-haired youth with greenish skin, entirely idiotic ... moving nothing but his black eyes and looking absolutely non-human.
That shape am I, I felt, potentially."

The narrator was left with a "horrible dread at the pit of my
stomach, and with a sense of insecurity of life that I never knew
before, and that I have never felt since." James attributed the account to an anonymous Frenchman; only after his book was published did he admit that the awful experience had been his.
Throughout his life James struggled with the dread and melancholy awakened in him by his glimpse of the abyss. He tried to
quell these "morbid feelings" with electrotherapy, hydrotherapy,
drugs, hypnotism, Christian Science, weightlifting, and various
"mind cures" that sought to harness the power of positive thinking—all apparently in vain.

Brian Josephson's Junction

The perils of mystical experience are evident in the tale of Brian Josephson, one of the most important and intriguing figures of twentieth-century physics. In 1962, when he was just a twenty-two-year-old graduate student at the University of Cambridge, he proposed that a special type of superconducting circuit, now known as a Josephson junction, should exhibit a seemingly magical quantum property now known as the Josephson effect. Josephson junctions are the basis of superconducting quantum interference devices (squids); these ultrasensitive instruments measure phenomena ranging from the whispers of neurons in human brains to the seismic mumbles of the earth. Josephson won a Nobel prize in 1973; he is one of the youngest recipients of the prize.

Shortly thereafter Josephson, who was already a tenured professor at Cambridge, renounced conventional physics and dedicated himself to the study of psychic and mystical phenomena and other forbidden matters. He began writing articles with titles like "Physics and Spirituality: The Next Grand Unification?" His contributions to mainstream journals consisted, for the most part, of letters denouncing science's narrow-minded attitude toward extrasensory perception and religion. In 1993, he contended in a letter to *Nature* that the religious impulse can help societies "function more harmoniously and more efficiently." He also proposed that religious practices stem from "genes linked to the potential for goodness." (Other letter writers promptly retorted that religions propagate intolerance and brutality at least as often as goodness.)

For years, I had heard physicists trade rumors about Josephson's metamorphosis. What happened? How could someone with so much scientific talent defect to the dark side? I had an opportunity to find out on the second day of the Tucson conference, when Josephson agreed to have lunch with me. Josephson looked as though he was trying to conceal his identity. His face was almost entirely concealed by his floppy white hat, thick black spectacles, shaggy hair, and sideburns. He wore a black T-shirt bearing the digitized portrait of Alan Turing, another British prodigy whose relations with the scientific establishment were troubled (although for very different reasons).

As we consumed burritos at a Taco Bell—surrounded by noisy Tucsonites, all of whom seemed young and tan and blond, especially compared to me and Josephson—Josephson told me, in a halting, low voice, about his past. He had begun to turn away from conventional physics in the mid-1960s. Like many other physicists, he became entranced by the seemingly crucial role of the observer in quantum mechanics and by the strange, nonlocal correlations linking inhabitants of the quantum realm. He was drawn to the works of sages such as Krishnamurti, an Indian mystic whose books cast a spell over many Western scientists and intellectuals in the 1960s. In 1966, while visiting the United States, Josephson befriended a mathematician with a strong interest in paranormal phenomena.

After some hemming and hawing, Josephson revealed that his transformation also sprang from changes "within." I asked him to elaborate. Did he have mystical or psychic experiences himself? "Well, in some ways, but not . . ." He paused. "I've had some strange experiences . . ." He prodded his burrito. Eventually he told me that in the late 1960s he fell into "hallucinatory states" as a result of working too hard on a physics problem. "My experiences were basically a result of a long period of having very little sleep," he said. For several years he took "major tranquilizers" to cope with his mental distress.

He managed to quell his inner turmoil through transcendental meditation. "Meditation provided enough stability where I didn't need" tranquilizers, he said. Josephson still meditated for up to several hours a day; the practice had given him "something like inner peace." His marriage in 1976 provided another anchor. He and his wife had a daughter, who was already showing talent as a writer. Discussing his daughter, Josephson permitted himself a rare smile.

He recalled feeling neither great joy nor great distress when he learned that he had received the Nobel prize in 1973. "I tend to be somewhat unemotional," he explained. "Mainly it was a nuisance, the amount of attention I got." On the other hand, the award gave him the confidence and opportunity to discuss publicly his interests in the more esoteric aspects of the mind. In his lectures and articles, Josephson scolded the scientific community for refusing to consider psychic phenomena, or "psi." He insisted that the data supporting telekinesis and extrasensory perception are "fairly convincing."

Quantum mechanics could help to account for ESP, Josephson asserted, but only if its scope is expanded. The current theory "doesn't allow the language of process or intention and so on. So I think we're going to have to extend quantum theory so we take that into account as well." Josephson added that "what is probably not properly addressed by science at the moment is the way parts form into wholes." Current reductionist methods were "excluding the possibilities that lie beyond the scope of that sort of a description." Josephson felt some kinship with David Bohm, a physicist who also advocated a holistic approach to science. (Bohm, in an interview shortly before his death in 1992, told me that he did not share Josephson's belief or interest in paranormal phenomena.)

Josephson had no regrets about having abandoned conventional physics. "I consider what I'm doing now to be more important." He had become accustomed to dealing with disapproval from other physicists and from officials at Cambridge. "It's not as bad as it used to be," he said. He occasionally arranged lectures on psychic phenomena in Cambridge, "and people on the whole have been quite impressed." Josephson only wished that funding agencies were enlightened enough to support his goal of forming a psi study group at Cambridge.

Josephson believed that meditation could help scientists enhance their abilities and insights. Ordinary consciousness, he explained, is "egoic." The ego "dominates everything," and one is no longer open to the influences and intuitions available to a "pre-egoic" child. Through meditation one could achieve a "trans-egoic" stage, in which "you gain the benefits of the processes that you were influenced by before the ego became dominant, while retaining some of the organizing ability of the ego."

That brought us, finally, to Josephson's theory of music, which he intended to discuss in his lecture here in Tucson. "Through my meditation I was able to sort of hear the music more deeply and see there was more in it," he explained. He came to believe that music stemmed, to some extent, not from superficial cultural influences but from timeless, universal "structures" of the mind. Scientists might learn something about these universal mental structures, Josephson suggested, by studying the human response to music. "So my intuition is that that may have great significance for our understanding of mind," he said.

Josephson's own tastes in music included classical and even a bit of rock and roll. "Some of that has considerable merit," he said of rock. "Something that may appear quite noisy, sometimes you get the feeling there is something quite deep to it." Did he have personal favorites? I asked. Josephson pursed his lips for a moment. He liked Simon and Garfunkel's "Bridge over Troubled Water," he said. "I don't know if that's particularly deep, but . . ."

In the background, the pop diva Whitney Houston was shrieking, "I'll always love youuuuuu!" The Taco Bell lunch throng had come and gone. Josephson finished off his burrito and taco, which he pronounced "quite good." He glanced at his watch; he was keen to get back to the conference to hear a lecture on "information physics, neuromolecular computing, and consciousness" by a scientist from Yugoslavia. We dumped our garbage in a wastebasket, placed our trays on a stack, and headed back out into the blinding day.

THE FUTURE OF
MIND-SCIENCE

. . . .

Anybody who has been seriously engaged in scientific work
of any kind realizes that over the entrance to the gates of the
temple of science are written the words: Ye must have faith.
It is a quality which the scientist cannot dispense with.

—MAX PLANCK

Mind-scientists have their creeds, just as religious believers do. Francis Crick spelled out his reductionist vision at the beginning of his book *The Astonishing Hypothesis.* " 'You,' your joys and your sorrows, your memories and your ambitions, your sense of personal identity and free will, are in fact no more than the behavior of a vast assembly of nerve cells and their associated molecules. As Lewis Carroll's Alice might have phrased it, 'You're nothing but a pack of neurons.' "

In a sense, Crick is right. We *are* nothing but a pack of neurons. At the same time, neuroscience has so far proved to be oddly unsatisfactory. Explaining the mind in terms of neurons has not yielded much more insight or benefit than explaining the mind in terms of quarks and electrons. There are many alternative reductionisms. We are nothing but a pack of idiosyncratic genes. We are nothing but a pack of adaptations sculpted by natural selection. We are nothing but a pack of computational devices dedicated to different tasks. We are nothing but a pack of sexual neuroses. These proclamations, like Crick's, are all defensible, and they are all inadequate.

In "More Is Different," an essay published in *Science* in 1972, Philip Anderson, a condensed-matter physicist at Princeton, brooded over the limits of scientific reductionism. Anderson had been piqued into writing the essay by the claim of particle physicists that they were performing the most fundamental—and thus most important—scientific research; everything else in science was merely "details" or, even worse, "engineering."

Anderson, who won a Nobel prize in 1977, acknowledged reductionism's extraordinary successes. Reductionism "is accepted without question" by the great majority of active scientists, he said. "The workings of our minds and bodies, and of all the animate or inanimate matter of which we have any detailed knowledge, are assumed to be controlled by the same set of fundamental laws." Nuclear physics, which addresses the smallest scale of reality, has provided insights into stars, galaxies, and the birth of the entire universe. Molecular biology, inaugurated by the discovery of the double helix, turned out to be an extraordinarily powerful approach to understanding evolution, heredity, embryonic development, and other aspects of life.

But knowledge of the basic laws governing the physical realm, Anderson pointed out, provides little illumination into many phenomena. Particle physics cannot predict the behavior of water, let alone the behavior of humans. Reality has a hierarchical structure, Anderson contended, with each level independent, to some degree, of the levels above and below. "At each stage, entirely new laws, concepts, and generalizations are necessary, requiring inspiration and creativity to just as great a degree as in the previous one," Anderson argued. "Psychology is not applied biology, nor is biology applied chemistry." If there is any feature of nature that has proved to be more than the sum of its parts, it is human nature.

The Myth of the Scientific Savior

Some mind-scientists, while acknowledging the limitations of all current approaches to the mind, prophesy the coming of a genius who will see patterns and solutions that have eluded all his or her predecessors. "It has happened," the Harvard psychologist Howard Gardner said to me. "It *will* happen." In his own lifetime, Gardner had witnessed the emergence of figures such as Noam Chomsky and Jean Piaget. "They said really profound things about the mind," Gardner elaborated. "They weren't necessarily right, but they certainly advanced the cause."

One possibility, Gardner suggested, will be that someone finds deep and fruitful commonalities between Western views of the mind and those incorporated into the philosophy and religion of the Far East. But Gardner emphasized that "we can't anticipate the extraordinary mind, because it always comes from a funny place that puts things together in a funny kind of way."

I heard much the same prediction from Eric Kandel, the Columbia neuroscientist. He noted that some philosophers whom he admired, such as Thomas Nagel of New York University, suspected that certain mind-related questions would never be solved. But Kandel had faith in the human mind to produce breakthroughs just when the situation seems bleakest. "There is an occasional person who will have a remarkable insight, that will allow you to see things in a new way, and that will move the field in unexpected directions."

But just how realistic is this myth of the scientific savior? In *Genius*, his biography of the physicist Richard Feynman, the science writer James Gleick addressed the widespread perception that contemporary culture no longer produces geniuses as towering as Newton or Mozart or Michelangelo. Gleick quoted the novelist Norman Mailer lamenting that "there are no large people any more. I've been studying Picasso lately and look at who his contemporaries were: Freud, Einstein."

Mailer's perception is an illusion, according to Gleick. In fact,

Gleick argued, there are so *many* Einsteins and Freuds alive to-day—so many brilliant scientists—that it has become harder for any individual to stand apart from the pack. This same reasoning explains why it has become harder for baseball players to attain a .400 batting average. (Of course, anomalies still occur, such as the seventy home runs that Mark McGwire hit in 1998.) Gleick's explanation seems sound to me, but I would add a crucial corollary: the scientific geniuses of our era have less to discover than their predecessors did. No modern scientist can discover gravity or natural selection or general relativity, because Newton and Darwin and Einstein got there first. To put things crudely, they solved the easy problems. The important problems that are left are extremely difficult.

That is not to say that geniuses cannot still have an impact. During the 1950s, particle physics was mired in a crisis that in some ways resembled the plight of neuroscience. Accelerators seemed to generate an exotic new particle almost daily; theorists had no idea how to organize the welter of findings into a cohesive theory. Then a brilliant young theorist named Murray Gell-Mann created a framework—which he jokingly called the Eight-Fold Way, after the Buddhist program for enlightenment—that categorized the particles according to shared properties. Later Gell-Mann and another physicist independently showed that many of these different particles were made of more fundamental particles called quarks.

But in terms of sheer complexity, particle physics is a child's game—a ten-piece jigsaw puzzle of *Snow White*—compared to neuroscience. Freud's ability to construct a unified theory of human nature was in large part a function of science's ignorance during his era. Anyone hoping to construct a unified theory of the mind now must cope with an astronomical number of findings, many of them with contradictory implications. When it comes to the human brain, there may *be* no unifying insight that transforms chaos into order.

The Dangers of Faith

Scientists will never accept that the mind cannot be tamed. Nor should they. It is always possible that they will find not only better remedies for mental illness but cures. They will learn how nature and nurture interact to produce not only human nature but an individual human. They will understand precisely how natural selection shaped and continues to constrain our minds. They will build machines that equal and surpass us in intelligence. They will solve the mind-body problem and the Humpty Dumpty dilemma. These outcomes are inevitable, optimists believe, given the steady and even precipitous pace of discovery and innovation in neuroscience, psychiatry, artificial intelligence, and other fields. All that is needed is sustained effort, funding—and a little faith.

But sometimes time, money, and faith are not enough to achieve even apparently reasonable scientific goals. The attempt to harness nuclear fusion, the process that makes the sun and other stars shine, is a case in point. The basic principles underlying fusion were known by the 1930s; physicists designed bombs based on those principles by the late 1940s. Given sufficient time and money, physicists would surely learn how to build fusion reactors that would generate energy much more cheaply and cleanly than dirty, expensive fission reactors. That vision never materialized. Even die-hard fusion enthusiasts are beginning to recognize that their dreams will probably never be realized; the technical, economic, and political obstacles to fusion energy are simply too great to overcome.

Cancer research provides what is perhaps a more appropriate analogy to mind-science. Unlike fusion energy, a cure for cancer is so compelling a goal that we are unlikely ever to abandon it. But so far a cure for cancer has proved to be just as elusive as fusion energy. Since President Richard Nixon officially declared a "war on cancer" in 1971, the United States has spent more than $35 billion on cancer research. Scientists have taken enormous strides toward understanding how different types of cancer occur, and they have invented sophisticated methods for detecting the disease and tracking its course. Certain rare childhood cancers have become more treatable, and even curable. But in spite of a recent, widely publicized decline in overall cancer mortality rates, the rates are *higher* now than they were in 1971, even when adjusted for the aging of the population.

There is something noble, even sublime, about the faith that propels scientists forward even after repeated failures. But this faith poses dangers, too. Elliot Valenstein addressed this issue at the end of *Great and Desperate Cures*, his history of the heyday of lobotomies in American psychiatry. The uncritical acceptance of lobotomies, far from being an aberration peculiar to psychiatry, is an all-too-common occurrence in modern medicine, Valenstein remarked. He cited recent instances in which scientists and the media had hailed untested treatments for AIDS, heart disease, Alzheimer's disease, and other illnesses:

> In the great majority of cases—where results are presented prematurely, where success is overestimated and dangers underestimated, where there are biases in the selection of patients, and where failures are explained away as exceptions—the physicians responsible have been convinced of the validity of their conclusions. Self-delusion is by its very nature difficult to guard against—almost impossible when fueled by unbridled ambition.

The dangers posed by scientific hubris are greatest when scientists seek not merely a cure for cancer or mental illness but a final, definitive explanation of who we are and even who we should be. The late philosopher Isaiah Berlin warned that applying science and reason to human affairs too often leads to totalitarianism. "A sense of symmetry and regularity, and a gift for rigorous deduction, that are prerequisites of aptitude for some natural sciences, will, in the field of social organization, unless they are modified by a great deal of sensibility, understanding and humanity, inevitably lead to appalling bullying on one side and untold suffering on the other." Berlin urged us to beware the "terrible simplifiers," "great despotic organizers," "men possessed by an all-embracing vision."

Of course, it is our own desire for answers and panaceas that gives the terrible simplifiers their power. To protect ourselves against our will to believe, we need to change the way we think and talk about mind-science. We need to remind ourselves how often mind-science has misled us in the past, and how little it has actually accomplished, while remaining open to the possibility of genuine advances. This is what I meant in the introduction to this book by the term "hopeful skepticism."

Howard Gardner, Clifford Geertz, and others have recommended that mind-science be viewed as a quasi-literary rather than strictly scientific enterprise. An exemplar of this literary approach is the neurologist and author Oliver Sacks. Sacks is the modern master of what I referred to in Chapter 1 as Gagian neuroscience. In his books and articles, he has provided extraordinarily vivid, empathetic profiles of people afflicted by autism, strokes, tumors, Tourette's syndrome, and other neurological disorders.

While most neuroscientists try to work around the irreducibility of each individual, Sacks has made it the centerpiece of his work. The poet William Carlos Williams proclaimed "no ideas but in things," violating the precept in stating it. Sacks's philosophy might be described as "no ideas but in people." Sacks once told me that he tried to follow Wittgenstein's precept that a book should consist of "examples" rather than generalizations. "People keep saying, 'Sacks, where's your general theory?' But I'm rather content to multipy case histories and leave the theorizing to others."

Sacks's compassionate, antireductionist credo is implicit within everything he writes, but occasionally he makes it explicit. In *The Man Who Mistook His Wife for a Hat*, Sacks wrote: "To restore the human subject at the centre—the suffering, afflicted, fighting, human subject—we must deepen a case history to a narrative or tale; only then do we have a 'who' as well as a 'what,' a real person, a patient, in relation to disease—in relation to the physical." In *An Anthropologist on Mars* he commented:

The realities of patients, the ways in which they and their brains construct their own worlds, cannot be comprehended wholly from observation of behavior, from the outside. In addition to the objective approach of the scientist, the naturalist, we must employ an intersubjective approach too, leaping, as Foucault writes, "into the interior of morbid consciousness, [trying] to see the pathological world with the eyes of the patient himself."

The problem with case histories is that while they often make compelling reading, they can obfuscate and subvert the truth. The case of Phineas Gage—the nineteenth-century man whose brain was pierced by an iron rod—demonstrated as much. The master of the case history was Sigmund Freud, who constructed psychoanalysis on cases such as Anna O., the Rat Man, the Wolf Man, and others. Scholars have shown that Freud's narratives often diverged sharply from the truth. Case histories have also provided distorted views of Prozac and other psychiatric drugs, of the links between genes and personality, and even of the role that natural selection plays in motivating human behavior.

Moreover, the vast majority of mind-scientists have neither the talent nor the inclination to present their results in a literary mode. Perhaps they should consider themselves engineers, as much so as bridge builders and circuit designers and automobile manufacturers. Engineers do not search for The Answer, the absolute, final, definitive Truth; thinking in such terms can even be an impediment to progress. Engineers search, rather, for *an* answer, for anything that helps to solve or ameliorate the problem at hand. By adopting such a humble stance, mind-science might acquire the same qualities that Prozac is supposed (erroneously) to possess: greatly increased benefits and minimal side effects.

Searching for an Epiphany

Ultimately the future of mind-science belongs to the young, and who knows where they will take it? In 1998, officials at the

Massachusetts Institute of Technology asked me to serve as a judge for a student essay contest. The students were asked to read two books—*Science: The Endless Frontier*, a paean to science's bottomless bounty written in 1945 by the physicist Vannevar Bush, and my gloomy tract, *The End of Science*—and then to set forth their own views of science's future. The essays were for the most part almost scarily well informed, articulate, and thoughtful. Many students singled out mind-science as a particularly promising area of research, but they also acknowledged potential limiting factors. "I have faith in an imminent cognitive revolution," one essayist proclaimed, but he warned that researchers could be stymied by taking too narrow and mechanistic a view of human nature. Another writer feared that artificial intelligence could be blocked by both the limits of silicon chip technology and a Luddite backlash.

My favorite essay wove together musings about cosmology, artificial intelligence, theology, and the essayist's beautiful but selfish former girlfriend. The author concluded with a prediction that "the future of science lies in mind-altering substances." He quoted from the British author Aldous Huxley, who after ingesting the psychedelic drug mescaline in the mid-1950s declared that such experiences "cannot be ignored by anyone who is honestly trying to understand the world in which he lives." (Although I nominated this essay for a prize, the other judges overruled me.)

My own sojourns into altered states have left me convinced that they cannot solve the mystery of consciousness. Far from it. I suspect that the more intelligent or aware or enlightened we become—whether through drugs or meditation or genetic engineering or artificial intelligence—the more we will be astonished, awestruck, dumbfounded by consciousness, and life, and the whole universe, regardless of the power of our scientific explanations. Wittgenstein captured this notion when he wrote, "Not *how* the world is, is the mystical, but *that* it is."

That is not to say that I don't still yearn for the epiphany that will make sense of everything. I briefly teetered on the verge of such a revelation at the consciousness conference in Tucson in 1994. It was my last night at the meeting, and I was consuming burritos and beers at an open-air restaurant with a half-dozen other conference goers, most of them science writers like me. Although the day had been blazingly hot, the night was cool. Discussing the meeting, we concurred that no one really knew what he or she was talking about; the scientists and philosophers were all lost and confused.

Some lectures were more interesting than others, of course. A favorite was Andrew Weil's tale of his toad-smoking exploits. My dinner partners seemed to agree with the alternative-medicine guru that consciousness would only be truly understood not from the outside but from the inside, not through science but through *experience*. We began trading stories about our own encounters with exotic mind-expanding substances: LSD, magic mushrooms, mescaline, peyote. A journalist who wore a chin braid and nose ring assured us that ketamine, sometimes called vitamin K, delivered the most mind-blowing trips of all. Ketamine was the drug that had enabled the neuroscientist John Lilly—pioneer of dolphin research and sensory-deprivation methods—to discover the extraterrestrial Beings who control our reality. Lilly described the Beings as solid-state machines who inhabit a dimensionless hyperspace consisting of pure consciousness and are concerned about humanity's maltreatment of dolphins and other animals.

As our conversation unraveled, a tall, moustached man wearing a collarless shirt splashed with blue flowers approached the table. He was carrying a contraption that consisted of goggles and headphones. He called it VARS, for Visual/Auditory Relaxation and Sedation. The man identified himself as a physician at the University of Arizona Health Sciences Center. He and a group of colleagues had invented the device and were testing its ability to soothe patients in physical or psychological distress. Promotional literature that I saw later described the gadget as a "non-invasive, non-pharmacological means of inducing relaxation and/or a hypnogogic state.... VARS employs the use of a programmable pulse generator that pulsates signals to an audio headphone and LED [light-emitting diode] fitted eyepieces. Synchronized visual and auditory stimulation (flashing lights and pulsating tone) is de-livered to the patient at varying frequencies."

When he asked if anyone would like to try it, I volunteered. Af-ter helping me pull the headphones over my ears and the goggles over my eyes, the man turned on a switch. Globules of sound and color rushed at me, welling up from subterranean depths. The tones swooped up and down, and the colors kept changing too, from red to blue to purple to yellow and back to red again. The sounds and colors merged; they became in some sense indistin-guishable, two aspects of the same essential sensation.

I heard voices, faint laughter, but they seemed to come from far away, from another world, another dimension. I focused only on these elemental sensations in my head, pulsing and transmuting, like the jewel of creation, ever changing and never changing, inde-scribably beautiful. I was looking into the heart of conscious-ness—not just my consciousness but all consciousness. The key to everything was there, waiting to be found, if I just looked hard enough. I felt an epiphany coming, a great revelation that would make everything clear. "Take a photograph of him and send it to his boss at *Scientific American*!" someone shouted, followed by hoots and guffaws. I realized that my mouth was open, and closed it. Slowly, reluctantly, I took off the goggles and headphones and reentered the world.

NOTES

PAGE   Introduction: I-WITNESSING

1   *"The sciences have developed"*: The quotation from Bertrand Russell is in *Quotationary*, edited by Leonard Frank, Random House, New York, 1999, p. 756. Frank cited Russell's 1935 book, *Religion and Science*.

1   *"I-witnessing"*: *Works and Lives*, by Clifford Geertz, Stanford University Press, Stanford, 1988, p. 79.

2   *Here, as elsewhere, I took my cue:* Stent's two books on the limits of science were *The Coming of the Golden Age*, Natural History Press, Garden City, New York, 1969, and *Paradoxes of Progress*, Freeman, San Francisco, 1978. See also my section on Stent in *The End of Science*, Broadway Books, New York, 1996, pp. 9–15.

3   *"the brain may not be capable"*: Stent, *Golden Age*, p. 74.

3   *Lewis Wolpert:* My encounter with Wolpert occurred at the London School of Economics on May 8, 1997, at a reception following a lecture by the evolutionary biologist John Maynard Smith. After Wolpert berated me, I was approached by an enormous bearded man, wearing a T-shirt and blue jeans, who introduced himself as Geoff Carr, the science editor of the *Economist*. He told me that he could not understand why some people were so upset by my book. Even if science ends, Carr quipped, we still have sex and beer. On May 19, after I had left England, the *Evening Standard*, a London newspaper, ran a review in which Wolpert stated his objections to *The End of Science* in a slightly more sober fashion. Wolpert wrote, "The very chapter title The End of Neuroscience makes a claim so silly that it is almost funny. Horgan interviews a few leading scientists like Edelman, Crick and Penrose, but the discussion of what we really understand about brain research remains superficial— and the scientists totally disagree with one another. This emphasizes nicely how far we have to go."

4   *"I am more concerned"*: *How We Die*, by Sherwin Nuland, Knopf, New York, 1993, p. 263.

5   *more than 1.2 billion people:* World Health Organization, press release, August 23, 1996.

5   *annual costs of brain-related disorders: Unlocking the Secrets of the Brain,* pamphlet issued to the press by American Psychiatric Association at its annual meeting in 1996.

6   *The evolutionary biologist Ernst Mayr:* Ernst Mayr discussed the limits of biology in *Toward a New Philosophy of Biology,* Harvard University Press, Cambridge, 1988.

6   *Thomas Kuhn contended:* Thomas Kuhn set forth his view of science in *The Structure of Scientific Revolutions,* University of Chicago Press, Chicago, 1962. See also my discussion of Kuhn and his work in *The End of Science,* pp. 41–47.

7   *"Paradigms, wholly new ways":* "Learning with Bruner," by Clifford Geertz, *New York Review of Books,* April 10, 1997, p. 22.

7   *"treatment of the id by the odd":* This quip, attributed to Macdonald Critchley, is in *Freudian Fraud,* by E. Fuller Torrey, HarperCollins, New York, 1992, p. xvi.

9   *Hippocrates hypothesized: Brain, Vision, Memory: Tales in the History of Neuroscience,* by Charles Gross, MIT Press, Cambridge, 1998, p. 12.

9   *"It is ... not possible": The Unnatural Nature of Science,* by Lewis Wolpert, Harvard University Press, Cambridge, 1993, p. 134.

11 *question-and-answer period:* This exchange took place when I lectured at the California Institute of Technology on May 18, 1998.

13 *I argued that Popper's scheme:* See *The End of Science,* pp. 33–41.

PAGE   Chapter 1: NEUROSCIENCE'S EXPLANATORY GAP

15 *"By 1979 Freudian psychology":* The quotation from Tom Wolfe is in "Post-Freudian Dream Therapy," by Martin Gardner, *Skeptical Inquirer,* November–December 1995, p. 56. Gardner cited Wolfe's book of essays *In Our Time,* Farrar, Strauss & Giroux, New York, 1980.

15 *"as the bones are lifted":* The quotation from Plato is in *Brain, Vision, Memory: Tales in the History of Neuroscience,* by Charles Gross, MIT Press, Cambridge, 1998, p. 16.

16 *"Materialism and Qualia":* "Materialism and Qualia: The Explanatory Gap," by Joseph Levine, *Pacific Philosophical Quarterly* 64, 1983, pp. 354–361.

16 *Society for Neuroscience:* One can obtain information on the Society for Neuroscience at its Web site: www.sfn.org.

16 *Nature Neuroscience:* The quotation is from a press release e-mailed to journalists on April 14, 1998.

17 *I once asked Gerald Fischbach:* I interviewed Gerald Fischbach at Harvard University on November 18, 1997.

18 *"Leaving something out": Consciousness Explained,* by Daniel Dennett, Little, Brown, Boston, 1991, p. 454.

19   *When I interviewed him:* I interviewed Torsten Wiesel at Rockefeller University on November 12, 1997.

19   *Nevertheless, in 1998 behavioral scientists:* "Next, the Decade of Behavior?" *Science,* January 16, 1998, p. 311.

19   *In 1958 Wiesel:* My account of the experiment by Wiesel and Hubel is based on "President Torsten Wiesel," by Geoffrey Montgomery, *Search* (a magazine published by Rockefeller University), Spring 1992, pp. 9–11.

20   *The researcher Karl Lashley:* An account of the work by Karl Lashley and others on memory can be found in *In the Palaces of Memory,* by George Johnson, Vintage Books, New York, 1992. For an excellent overview of memory research, see also *Searching for Memory,* by Daniel Schacter, Basic Books, New York, 1996.

21   *Karl Friston, an MRI specialist:* I interviewed Karl Friston by telephone on April 20, 1998.

22   *Rodolfo Llinas, a neuroscientist:* I interviewed Rodolfo Llinas at New York University on April 28, 1998.

23   *"This surprising tendency":* Eye, Brain and Vision, by David Hubel, Scientific American Library, New York, 1988, p. 220.

23   *Patricia Goldman-Rakic, a professor:* I interviewed Patricia Goldman-Rakic at Yale University on December 19, 1997. For an account by Goldman-Rakic of her own work, see "Working Memory and the Mind," *Scientific American,* September 1992, pp. 111–117.

28   *I edited an article:* "The Machinery of Thought," by Timothy Beardsley, *Scientific American,* August 1997, pp. 78–83. The letter responding to the article was published in the December 1997 issue, p. 8. Upon reading this section of the chapter in draft, Chris Bremser, a prominent San Francisco–based computer scientist, commented, "Anyone remotely involved with computer programming will recognize this explanatory gap as being exactly analogous to the difference between machine code (1's and 0's) and actual programs. Anyone can hook a scope up to a computer's memory, but without having (or inferring) the higher-level programmatic constructs at work, no insight will be gained. Douglas Hofstadter discusses this at almost infinite length in *Gödel, Escher, Bach,* Vintage Books, New York, 1980."

29   *Cognitive science "is really":* The Emotional Brain, by Joseph LeDoux, Simon & Schuster, New York, 1996, p. 25.

29   *LeDoux, himself a cool, controlled man:* I interviewed Joseph LeDoux at New York University on February 6, 1998.

29   *LeDoux and his colleagues showed:* The description of LeDoux's work is in Schacter, *Searching for Memory,* p. 214.

30   *"are red herrings":* LeDoux, *The Emotional Brain,* p. 18.

31   *"We have no idea":* LeDoux made these remarks in response to a posting by me on an Internet site called "The Edge," www.edge.org.

32   *Gage "talked so rationally":* The quotation is in *Descartes's Error*, by Antonio Damasio, Avon Books, New York, 1994, p. 6. My account of Phineas Gage is drawn primarily from ibid. and *An Anthropologist on Mars*, by Oliver Sacks, Vintage Books, New York, 1995.

33   *"completely recovered":* Ibid., p. 60.

33   *"fitful, irreverent":* Ibid., p. 61.

34   *A Swiss political journalist:* "Is Everybody Crazy?" by Sharon Begley, *Newsweek*, January 26, 1998, p. 52.

34   *A slew of self-help books:* The examples cited in this paragraph are from *The Right Mind*, by Robert Ornstein, Harcourt Brace, New York, 1997, pp. 87–96.

34   *a critique of the welfare state:* Ibid., pp. 90–91. Michael Gazzaniga's book was *The Social Brain*, Basic Books, New York, 1985.

35   *In an article published in* Scientific American: "The Split Brain Revisited," by Michael Gazzaniga, *Scientific American*, July 1998, pp. 50–55.

35   *A British boy named Alex:* "Removing Half of Brain Improves Young Epileptics' Lives," by Abigail Zuger, *New York Times*, August 19, 1997.

36   *"Because every individual":* Last Resort, by Jack Pressman, Cambridge University Press, New York, 1998, p. 434.

36   *scientists from around the world:* "Penetrating Insight into the Brain," *Science*, October 2, 1998, p. 39.

36   *"In the training and in the exercise of medicine":* The undated quotation from Sherrington is in *Isaac Asimov's Book of Science and Nature Quotations*, edited by Isaac Asimov and Jason Shulman, Weidenfeld and Nicholson, New York, 1988, p. 228.

37   *widely publicized MRI study:* "Cerebral Anatomical Abnormalities in Monozygotic Twins Discordant for Schizophrenia," by Richard Suddath et al., *New England Journal of Medicine*, March 22, 1990. NIMH director Lewis Judd commented on the study in "Brain Structure Differences Linked to Schizophrenia in Study of Twins," by Daniel Goleman, *New York Times*, March 22, 1990, p. B15.

37   *"making fundamental discoveries":* Galen's Prophecy, by Jerome Kagan, Basic Books, New York, 1994, p. 274.

38   *article in* American Scientist: "Psychological Science at the Crossroads," by Richard Robins, Samuel Gosling, and Kenneth Craik, *American Scientist*, July–August 1998, pp. 310–313.

38   *"Anyone interested":* "Shrinking Minds and Swollen Heads," by V. S. Ramachandran and J. J. Smythies, *Nature*, April 17, 1997, pp. 667–668.

38   *"One of the reasons":* Susan Greenfield's remarks appeared in *On Giants' Shoulders*, by Melvyn Bragg, Hodder & Stoughton, London, 1998, pp. 235–236.

39   *When I asked Bloom:* I interviewed Floyd Bloom by telephone on March 11, 1998.

39   *Edelman remarked: Bright Air, Brilliant Fire*, by Gerald Edelman, Basic Books, New York, 1992, p. 145.

40 *brilliance and bullying:* See the portrait of Eric Kandel in Johnson, *In the Palaces of Memory*, pp. 59–63.

40 *two leading neuroscience textbooks: Principles of Neural Science,* edited by Eric Kandel and James Schwartz, Elsevier North-Holland, New York, 1981; *Essentials of Neural Science and Behavior,* edited by Eric Kandel, James Schwartz, and Thomas Jessell, Appleton & Lange, Stamford, Connecticut, 1995.

40 *and has exerted:* Kandel himself told me that he had influenced the coverage of neuroscience at *Scientific American* and the *New York Times.*

40 *"purplish-green baked potato with ears":* "Our Memories, Our Selves," by Stephen Hall, *New York Times Magazine,* February 15, 1998, p. 30.

41 *"e = mc² of the mind":* Ibid., p. 28.

41 *"pioneered much of the research":* Ibid., p. 28.

41 *"fruitfully distracted":* Ibid., p. 30.

41 *Kandel spelled out:* "A New Intellectual Framework for Psychiatry," by Eric Kandel, *American Journal of Psychiatry,* April 1998, pp. 457–469. See also "Psychotherapy and the Single Synapse," by Eric Kandel, *New England Journal of Medicine,* November 8, 1979, pp. 1028–1037.

42 *I met Kandel:* I interviewed Kandel at the New York Psychiatric Institute on December 15, 1997.

45 *Freud spent more than a decade:* The information about Freud's early career is in "Psychoanalysis and Neuroscience," a special issue of the *Journal of Clinical Psychoanalysis,* edited by Herbert Wyman and Stephen Rittenberg, Volume 5, Number 3, 1996; and in "The Other Road: Freud as Neurologist," by Oliver Sacks, in *Freud: Conflict and Culture,* edited by Michael Roth, Knopf, New York, 1998, pp. 221–234. This book served as the catalogue for the U.S. Library of Congress's exhibit on Freud.

45 *"One evening last week":* Ibid., p. 230.

45 *"The intention is":* The Freud Reader, edited by Peter Gay, Norton, New York, 1989, p. 87.

45 *"I no longer understand":* As quoted in Sacks, "The Other Road: Freud As Neurologist," p. 230.

46 *"We know two things":* As quoted in Wyman and Rittenberg, "Psychoanalysis and Neuroscience," p. 332.

PAGE Chapter 2: WHY FREUD ISN'T DEAD

47 *It is quite possible: Language and Problems of Knowledge,* by Noam Chomsky, MIT Press, Cambridge, 1988, p. 159.

47 *The occasion was a meeting:* The meeting of Division 39 of the American Psychological Association took place at the Waldorf-Astoria in New York City on April 18, 1996.

49 *"a scientific fairy tale":* Freudian Fraud, by E. Fuller Torrey, HarperCollins, New York, 1992, p. 216.

49   *"If the patient loved":* Ibid.

49   *"a total lack":* Ibid.

49   *"well founded neither theoretically":* Ibid.

49   *"the green cheese hypothesis":* Ibid.

49   a *"witchdoctor"* and *"Viennese quack":* Ibid., p. 200.

49   *Attacks on Freud: Why Freud Was Wrong,* by Richard Webster, Basic Books, New York, 1995; *Freud Evaluated,* by Malcolm Macmillan, MIT Press, Cambridge, 1997; *Unauthorized Freud,* edited by Frederick Crews, Viking, New York, 1998.

49   *the Loch Ness monster:* "The Freud Controversy," by Frank Cioffi, in *Freud: Culture and Conflict,* edited by Michael Roth, Knopf, New York, p. 181.

50   *as* Time *magazine did:* "Is Freud Dead?" by Paul Gray, *Time,* November 29, 1993, pp. 47–51.

50   *"Freud's recent critics": Freud and His Critics,* by Paul Robinson, University of California Press, Berkeley, 1993, p. 269.

50   *One survey of literature:* I found this fact in *Freud Scientifically Reappraised,* by Seymour Fisher and Roger Greenberg, John Wiley, New York, 1996, p. 8.

50   *"This does not mean":* "Psychological Science at the Crossroads," by Richard Robins, Samuel Gosling, and Kenneth Craik, *American Scientist,* July–August 1998, p. 311.

51   *Membership in:* The American Psychoanalytic Association supplied these figures about U.S. and international enrollment. For an upbeat report on the status of psychoanalysis, see also "Return to the Couch: A Revival for Analysis," by Erica Goode, *New York Times,* January 12, 1999, p. C1.

51   *In 1996 the Russian president:* "Freud in Russia: Return of the Repressed," by Alessandra Stanley, *New York Times,* December 11, 1996, p. A1.

51   *"No negative critique":* Webster, *Why Freud Was Wrong,* p. 455.

51   *neuroses are "without exception": The Freud Reader,* edited by Peter Gay, Norton, New York, 1989, p. 15.

52   *"frivolous and premature":* Ibid., p. 34.

52   *"The amount of effort":* Ibid., p. 18.

52   *"heads I win":* Ibid., p. 666.

52   *"Some Psychical Consequences":* Ibid., pp. 671–678.

53   *"Women have never fared well": Madness on the Couch,* by Edward Dolnick, Simon & Schuster, New York, 1998, p. 283.

53   *a Manhattan neighborhood:* Manhattan's "mental block" is West 81st Street between Central Park West and Amsterdam Avenue.

53   *"culture-bound and outmoded": The Talking Cure,* by Susan Vaughan, G. P. Putnam's Sons, New York, 1997, pp. 4–5.

54   *"Insight such as this":* Gay, *The Freud Reader,* p. 129.

55   *The* New York Times *ran a story:* "Was Freud Wrong? Are Dreams the Brain's Start-Up Test?" by Nicholas Wade, *New York Times,* January 6, 1998, p. F6. The letters responding to the story were printed on January 12.

55   *Empirical studies:* For a discussion of the Oedipal complex from an evolutionary point of view, see *How the Mind Works,* by Steven Pinker, Norton, New York, 1997, pp. 459–460.

55   *What is one to make:* "Mothers Determine Sexual Preferences," by Keith Kendrick et al., *Nature,* September 17, 1998, pp. 229–230.

56   *That became clear:* I interviewed Roger Greenberg by telephone on June 11, 1998.

57   *In his 1992 book:* E. Fuller Torrey's review of studies of the anal hypothesis can be found in *Freudian Fraud,* pp. 265–277. Seymour Fisher's study of the anal hypothesis is summarized on pp. 276–277.

58   *when he attacked Freud:* Crews's two articles were published on November 18, 1993, and December 1, 1994. The articles plus correspondence inspired by them were later published as a book: *The Memory Wars,* by Frederick Crews, New York Review of Books, New York, 1995.

58   *"dishonesty and cowardice":* Ibid., p. 59.

58   *"It is not recorded":* Ibid., p. 39.

59   *It has been estimated:* Ibid., p. 159.

60   *"Before they come":* Ibid., p. 58.

60   *"By virtue of his prodding":* Ibid., p. 72.

60   *One observer commented:* "The Bewildered Visionary," by Richard Webster, *Times Literary Supplement,* May 16, 1997, p. 10.

61   *I first encountered Crews:* I interviewed Crews on April 2, 1998, in New Haven, Connecticut.

62   *an Italian scholar:* Sebastiano Timpanaro presents his view of Freudian slips in "Error's Reign," *Unauthorized Freud,* pp. 94–105.

63   *Crews began his lecture:* "Whose Freud?" was held at Yale University April 3–4, 1998.

65   *In one packed session:* The five participants at the Division 39 session titled "Individual and Marital Treatment of an Incest Survivor," held at the Waldorf on April 18, 1996, were Lynn Passey, Marylou Lionells, Sue Grand, Darlene Bregman Ehrenberg, and Jody Messler Davies.

65   *a talk whose central theme:* The analyst's name was Philip Bromberg, and his talk was listed in the meeting program as "Staying Sane While Changing: Reflections on Clinical Judgement." Bromberg noted at the beginning of his talk that the program title was a misprint; it should have read "Staying the *Same* While Changing." But Bromberg said that the unintentional title, like a Freudian slip, conveyed his meaning better than the intentional title.

66 *Freud denied the "radical" claim:* Gay, *The Freud Reader*, p. 721.

67 *"superperfectionistic perspective":* Fisher and Greenberg, *Freud Scientifically Reappraised*, pp. 11–12.

67 *Shortly after his appointment:* I interviewed Hyman in New York City on May 7, 1996.

69 *"a genius not of science":* Torrey, *Freudian Fraud*, p. 218.

69 *"By modern standards": The Astonishing Hypothesis*, by Francis Crick, Charles Scribner's Sons, New York, 1994, p. 14.

70 *a form of "shamanism": The Western Canon*, by Harold Bloom, Harcourt Brace, New York, 1994, pp. 376–377.

70 *"For all of his log rolling":* Gray, "Is Freud Dead?" p. 51.

70 *Frederick Crews once fell:* The information about Crews's Freudian past is in an excellent profile of Crews, "Terminating Analysis," by Adam Begley, *Lingua Franca*, July–August 1994, p. 28.

71 *Geertz's term* faction: *Works and Lives*, by Clifford Geertz, Stanford University Press, 1988, p. 141.

71 *his 1983 book: Frames of Mind*, by Howard Gardner, Basic Books, New York, 1983; *Extraordinary Minds*, by Howard Gardner, Basic Books, New York, 1997.

71 *I had expected:* I interviewed Howard Gardner at Harvard University on November 19, 1997.

72 *Gardner had first offered:* "Scientific Psychology: Should We Bury It or Praise It?" by Howard Gardner, *New Ideas in Psychology* 10, no. 2, 1992, pp. 179–190.

72 *"Psychology has not added":* Ibid., p. 180. Gardner also discussed the lack of progress in psychology in "Perennial Antinomies and Perpetual Redrawings: Is There Progress in the Study of the Mind?" in *Science of the Mind: 2001 and Beyond*, edited by R. Solso and D. Massaro, Oxford University Press, New York, 1995, pp. 65–78. Gardner stated on p. 67: "A dispassionate history of the last hundred years scarcely reveals a field that is making steady progress. It is at least as convincing to argue that psychology has been a succession of, and a struggle among, a number of rival schools and paradigms: functionalism, structuralism, behaviorism, psychoanalysis, Gestalt psychology, and most recently, information-processing, connectionist, and sociobiological approaches. . . . Some would contend that this struggle among schools is inevitable, the sign of a young and dynamic field, and would argue that there has been deep, underlying progress nonetheless. . . . For the most part, however, the schools have not succeeded so much as they have become exhausted: the names disappear but the struggles continue under new banners."

73 *his 1985 book: The Mind's New Science*, by Howard Gardner, Basic Books, New York, 1985.

74 *"At its best"*: "Paging Dr. Freud," by Adam Phillips, *New York Times Book Review*, June 7, 1998, p. 24.

74 *"It almost looks as if"*: Quoted on the title page of *Psychoanalysis: The Impossible Profession*, by Janet Malcolm, Vintage Books, New York, 1982.

74 *"I do not think"*: Fisher and Greenberg, *Freud Scientifically Reappraised*, p. 204.

PAGE Chapter 3: PSYCHOTHERAPY AND THE DODO HYPOTHESIS

75 "We've Had a Hundred Years": *We've Had a Hundred Years of Psychotherapy and the World's Getting Worse*, by James Hillman and Michael Ventura, Harper San Francisco, 1993.

76 *more than 450 different forms:* "The Psychotherapies: Benefits and Limitations," by Toksoz Karasu, *American Journal of Psychotherapy* 40, no. 3, July 1986, pp. 324–341.

76 *three broad categories:* These categories are adopted from "Psychotherapies: An Overview," by Toksoz Karasu, *American Journal of Psychiatry*, August 1977, pp. 851–863. See also *Am I Crazy, or Is It My Shrink?* by Larry Beutler et al., Oxford University Press, New York, 1998, p. 99. Beutler and his colleagues created a separate category for interpersonal therapies, which Karasu had categorized as a type of psychodynamic therapy. New psychotherapies emerge almost daily, of course. See "Philosophers Ponder a Therapy Gold Mine," by Joe Sharkey, *New York Times Week in Review*, March 8, 1998, p. 1. The article reported that philosophers are beginning to offer their services as therapists. See also "You Are Getting Very Confused: Psychologists' Split Decisions," *New York Times Week in Review*, June 14, 1998, p. 7. The article reported on a book titled *Escaping the Advice Trap*, by Wendy Williams and Stephen Ceci, Andrews McMeel, Kansas City, Missouri, 1998. The book revealed that different therapists, when presented with identical cases, arrived at very different diagnoses and offered very different advice to the patient. The authors, both psychologists at Cornell University, concluded that the solution to this problem is not to dispense with psychotherapists but to see more than one.

77 *anecdotal evidence:* See, for example, "The Final Analysis," by James Kaplan, *New York Magazine*, October 20, 1997, pp. 26–33.

77 *In 1987 73 million Americans:* "Outpatient Psychotherapy in the United States, I: Volume, Costs and User Characteristics," by Mark Olfson and Harold Alan Pincus, *American Journal of Psychiatry*, September 1994, p. 1284.

77 *80 million and cost more than $4 billion:* Ibid., p. 1281.

77 *A 1992 study counted:* "Price Tag: Psychotherapy," *New York Times*, Feb-

ruary 4, 1993, p. C1. The *Times* did not specify the source for this statistic.

77   *A federal survey focusing on physicians:* These data come from the annual National Ambulatory Medical Care Survey, which is conducted by the Centers for Disease Control and Prevention, National Center for Health Statistics. The total number of psychotherapy sessions conducted in physicians' offices fell from more than 22 million in 1989 to fewer than 16 million in 1996. David Woodwell of the Division of Health Care Statistics provided these data.

77   *The United States alone harbors:* I compiled these figures on different types of psychotherapists from interviews with the American Psychiatric Association, the American Psychological Association, and the National Association of Social Workers. The number of workers who are at least potentially qualified to practice psychotherapy or counseling of some kind is much larger. The National Bureau of Labor Statistics (Web site address: stats.bls.gov/oes/national/oes_prof.htm) lists the following numbers: 246,100 psychiatric and medical social workers, 240,220 human services workers, 196,530 residential counselors, 92,630 psychologists, and 25,360 members of the clergy. See also *Manufacturing Victims*, by Tana Dineen, Robert Davies Multimedia Publishing, Quebec, Canada, 1996. Dineen, a Canadian psychologist, reported that the percentage of Americans who have seen a "mental health professional" rose from 14 percent in the mid-1960s to 46 percent in 1995; the number of licensed psychologists per capita almost doubled from 1975 to 1995.

77   *The so-called:* For information on the Mental Health Parity Act, see "Insurance Plans Skirt Requirement on Mental Health," by Robert Pear, *New York Times*, December 26, 1998, p. A1.

78   *Case histories remain:* See "Emotional Displays," by Stuart Sutherland, *Nature*, December 4, 1997, p. 459. Sutherland, a British journalist and psychologist who suffered from manic depression and died in 1998, wrote: "Case histories can undoubtedly be a source of hypotheses, but to have validity the hypotheses need to be refined and tested by experiment. After all, Freud, relying on case histories, produced the most spectacularly wrong theory of the century."

79   *"Attempting to answer":* Karasu, "The Psychotherapies: Benefits and Limitations," p. 335.

79   *The American Psychiatric Association has sought: Diagnostic and Statistical Manual of Mental Disorders*, 4th ed., American Psychiatric Association, Washington, D.C., 1994.

79   *"boils down to this":* "The Encyclopedia of Insanity," by L. J. Davis, *Harper's*, February 1997, p. 65.

79   *official disorders surged:* The data are from "You're Not Bad, You're Sick. It's in the Book," by Joe Sharkey, *New York Times Week in Review*, September 28, 1997.

80   *"the growing tendency": Making Us Crazy*, by Herb Kutchins and Stuart Kirk, Free Press, New York, 1998, p. x.

81   *Large-scale statistical studies: Freud Scientifically Reappraised*, by Seymour Fisher and Roger Greenberg, John Wiley, New York, 1996, p. 195.

81   *"One ought not":* Ibid., p. 204.

81   *"out for bigger game": Psychoanalysis: The Impossible Profession*, by Janet Malcolm, Vintage Books, New York, 1982, p. 123.

81   *Freud himself once chided:* Ibid., p. 124.

81   *In 1948, leaders:* The description of the initial efforts of leaders of the American Psychoanalytic Association to gather evidence on the efficacy of psychoanalysis is in *A History of Psychiatry*, by Edward Shorter, John Wiley, New York, 1997, pp. 311–312.

82   *"anodyne factoids":* Ibid., p. 312.

82   *Nevertheless, in the 1950s:* I constructed this account of the study of Bachrach et al. from an interview with Bachrach in 1996 and from materials that he distributed at the Annual Meeting of the American Association for the Advancement of Science in Chicago, February 7, 1992. Bachrach and his colleagues also published their findings in the *Journal of the American Psychoanalytic Association* 39, no. 4, 1991, pp. 871–916. In *Freud Scientifically Reappraised*, p. 201, Fisher and Greenberg commented that "there is no study of psychoanalysis as a treatment that cannot be dismissed because of seriously compromised or contaminated data."

83   *Cambridge-Somerville Delinquency Prevention Project: Freudian Fraud*, by E. Fuller Torrey, HarperCollins, New York, 1992, pp. 168–169.

83   *research by Hans Eysenck:* Shorter, *A History of Psychiatry*, p. 312.

83   *Eysenck's overt hostility:* The background information on Eysenck is in Torrey, *Freudian Fraud*, p. 218, and in Eysenck's obituary in the *New York Times*, September 10, 1997, p. A27.

84   *"not entirely convincing": Toxic Psychiatry*, by Peter Breggin, St. Martin's Press, New York, 1991, p. 404.

84   *One of the most influential reports:* "Comparative Studies of Psychotherapies: Is It True That 'Everybody Has Won and All Must Have Prizes'?" by Lester Luborsky et al., *Archives of General Psychiatry* 32, 1975, pp. 995–1008. I interviewed Luborsky by telephone several times about his research.

85   *Rosenzweig had postulated:* "Some Implicit Common Factors in Diverse Methods of Psychotherapy," by Saul Rosenzweig, *American Journal of Orthopsychiatry* 6, 1936, pp. 412–415.

85   *"allegiance effect":* See "The Efficacy of Dynamic Therapies," by Luborsky et al., in *Psychodynamic Treatment Research*, New York, Basic Books, 1993, pp. 508–509.

86   *One exchange in the dialogue:* Ibid., p. 511.
86   *an elaborate hypothesis:* Luborsky's theory, called the core conflictual relationship theme, is explained in *Understanding Transference,* by Luborsky and Paul Crits-Christoph, American Psychological Association, Washington, D.C., 1998.
86   *Luborsky divulged:* Luborsky's letter to the *New York Review of Books* was reprinted in *The Memory Wars,* by Frederick Crews, New York Review of Books, New York, 1995, pp. 102–104. Crews's response is on pp. 129–130.
87   *The term* placebo: This background on the placebo can be found in the introduction of *The Placebo Effect,* edited by Anne Harrington, Harvard University Press, Cambridge, 1997. Harrington wrote the introduction.
87   *"until recently, the history":* "The Placebo: Is It Much Ado About Nothing?" by Arthur Shapiro and Elaine Shapiro, in ibid., p. 13.
87   *theriac:* Ibid., p. 14.
87   *Galen once wrote:* Ibid., p. 13. See also *The Powerful Placebo,* by Arthur Shapiro and Elaine Shapiro, Johns Hopkins University Press, Baltimore, 1997.
87   *"Hurry, hurry":* Ibid., p. 14. Leon Hoffman of the American Psychoanalytic Association suggested to me that this phenomenon, in which newer drugs evoke a stronger placebo effect, may be related to the infamous Hawthorne effect. The Hawthorne effect dates back to a study carried out at a Western Electric plant in Hawthorne, Illinois, from 1927 to 1933. It sought to determine whether changing the plant's décor, coffee break schedules, and other conditions would affect productivity. The results indicated that change, in and of itself, tended to boost the mood and productivity of workers. The "much embraced" study was actually based on only three subjects, according to "Scientific Myths That Are Too Good to Die," by Gina Kolata, *New York Times Week in Review,* December 6, 1998, p. 2.
88   *Henry Beecher revealed:* Harrington, *The Placebo Effect,* pp. 2–3.
88   *arterial ligation:* From "The Placebo Effect," by Walter Brown, *Scientific American,* January 1998, p. 92.
88   *The psychologist Robert Ader:* The rat experiments of the psychologist Robert Ader are described in Harrington, *The Placebo Effect,* p. 5.
88   *"power of the placebo" is reflected:* Ibid., p. 24.
89   *medieval potion theriac:* Ibid., p. 23.
89   *"To our astonishment":* Persuasion and Healing, by Jerome Frank and Julia Frank, Johns Hopkins University Press, 3rd ed., Baltimore, 1993, p. 298.
89   *"relief of anxiety and depression":* Ibid., p. 152.
90   *"The methods of both":* Ibid., p. 66.

90   *"resembles a text":* Ibid., p. 300.
90   *"faith in science still seems":* Ibid., p. 42.
91   *shamans and faith healers:* Ibid., pp. 87–112.
91   *Crews favored treatments:* Frederick Crews made these remarks when
       I interviewed him on April 2, 1998.
91   *Martin Seligman of the University of Pennsylvania:* Martin Seligman gave
       me his views on cognitive therapy during a telephone interview in
       1996.
91   *one variant of cognitive therapy:* Karasu, "Psychotherapies: An Over-
       view," p. 858.
92   *"He saved my life":* "Changing Thinking to Change Emotions," by
       Jane Brody, *New York Times*, August 21, 1996, p. C9.
92   *Shear and her colleagues concluded:* "Cognitive Behavioral Treatment
       Compared with Nonprescriptive Treatment of Panic Disorder," by
       M. Katherine Shear et al., *Archives of General Psychiatry* 51, 1994, pp.
       395–401.
92   *The Dodo hypothesis has:* "Meta-analysis of Psychotherapy Outcome
       Studies," by Mary Smith and Gene Glass, *American Psychologist* 32,
       1977, pp. 752–760. The study is described in *House of Cards*, by Robyn
       Dawes, Free Press, New York, 1994, p. 50.
93   *the pseudotherapists:* "Specific Versus Non-specific Factors in Psy-
       chotherapy," *Archives of General Psychiatry* 36, 1979, pp. 1125–1136, cited
       in Dawes, *House of Cards*, p. 56.
94   *"Those claiming to be":* Dawes, *House of Cards*, p. 5. I interviewed Robyn
       Dawes by telephone several times.
94   *I Want to Live!* Ibid., p. 146.
95   *"It is not clear to me":* This memo is reproduced in the preface of *The
       Case for Pragmatic Psychology*, by Daniel Fishman, New York Univer-
       sity Press, New York, 1999.
95   *"spend their time doing":* Torrey, *Freudian Fraud*, p. 251.
95   *"the sad spectacle":* Ibid.
96   *I toured a museum:* I visited the Hudson River Psychiatric Center on
       September 29, 1997. I am grateful to my friends Jan and Alan Peter-
       son for telling me about the center's museum.
100   *patients housed in state asylums:* The figures are in "Prisons Replace
       Mental Hospitals for the Nation's Mentally Ill," by Fox Butterfield,
       New York Times, March 5, 1998, p. A1.
100   *Torrey himself:* "The Release of the Mentally Ill from Institutions: A
       Well-Intentioned Disaster," by E. Fuller Torrey, *Chronicle of Higher
       Education*, June 13, 1997, pp. B4–5. Torrey and other advocates for the
       mentally ill stepped up their criticism of deinstitutionalization in
       the summer of 1998 after several highly publicized cases in which
       schizophrenics committed murder. See "Fearsome Madness," by
       Wray Herbert, *U.S. News & World Report*, August 10, 1998, pp. 53–54. In
       this story, Laurie Flynn of the National Alliance for the Mentally Ill

was quoted saying that it might be time to resurrect "the old notion of asylum, to protect the mentally ill as well as society."

PAGE   Chapter 4: PROZAC AND OTHER PLACEBOS

102   *"In time, I suspect"*: *Listening to Prozac*, by Peter Kramer, Penguin Books, New York, 1993, p. 300.

102   *In May 1996*: The American Psychiatric Association meeting took place in New York City on May 4–9, 1996.

105   *"If there is one intellectual reality"*: *A History of Psychiatry*, by Edward Shorter, John Wiley, New York, 1997, p. vii.

105   *"sounds relatively benign"*: "A Shrinking Discipline," by John Marchall, *Nature*, March 27, 1997, p. 346. For an even more scathing review of Shorter, see "Chlorpromazine Is No Penicillin," by Andrew Scull, *Times Literary Supplement*, May 16, 1997, pp. 8–9.

105   *Freud himself employed electrotherapy*: *The Freud Reader*, edited by Peter Gay, Norton, New York, 1989, p. 9.

106   *malaria, tuberculosis, typhoid*: Shorter, *A History of Psychiatry*, pp. 192, 247.

106   *"an epochal moment"*: Ibid., p. 194.

106   *displaced by insulin-coma therapy*: Ibid., pp. 209–213.

106   *"was never a big success"*: Ibid., p. 216.

107   *The "sleep cure"*: Ibid., pp. 200–207.

107   *sleep deprivation*: See "Drug Makers' Goal: Prozac Without the Lag," by Thomas Burton, *Wall Street Journal*, April 27, 1998, p. B1.

107   *"brainwashing"*: Shorter, *A History of Psychiatry*, p. 207.

107   *patients given apomorphine*: Ibid., p. 199.

107   *turpentine, sulfur, and other toxins*: Ibid., p. 247.

107   *laxatives*: Ibid., p. 196.

107   *colons, ovaries, gonads*: Ibid., p. 175.

108   *the "treatment of choice"*: Ibid., p. 223.

108   *"Georgia Power Cocktail"*: Ibid., p. 282.

108   *not a single nonwhite American*: "Electroconvulsive Therapy," by Harold Sackheim et al., *Psychopharmacology: The Fourth Generation of Progress*, edited by Floyd Bloom and David Kupfer, Raven Press, New York, 1995, p. 1123.

108   *"taught us to look with less awe"*: *The Three-Pound Universe*, by Judith Hooper and Dick Teresi, St. Martin's Press, New York, 1986, p. 40.

109   *In one five-week period*: *Great and Desperate Cures*, by Elliot Valenstein, Basic Books, New York, 1986, p. 229.

109   *"headhunting" expedition*: "Psychosurgery Redux," by Wray Herbert, *U.S. News & World Report*, November 3, 1997, p. 63.

109   *some forty thousand lobotomies*: Ibid.

109   *incorrigible prisoners*: *Last Resort*, by Jack Pressman, Cambridge University Press, New York, 1998, p. 406.

109   *Freeman reportedly carried out*: *Toxic Psychiatry*, by Peter Breggin, St.

Martin's Press, New York, 1991, pp. 31–32. Breggin claimed that Freeman gave him this estimate in a telephone conversation.

109 *twenty-five in a single day:* Valenstein, *Great and Desperate Cures,* p. 231.

109 *Freeman's last patient:* Ibid., p. 274.

110 *"garrulous, euphoric":* Shorter, *A History of Psychiatry,* p. 257.

110 *In 1955* Time *proclaimed:* Ibid., p. 254.

110 *"The future may teach us":* Quoted in "Is Freud Dead?" by Paul Gray, *Time,* November 29, 1993, p. 47.

111 *as many as 50 percent: From Placebo to Panacea,* edited by Seymour Fisher and Roger Greenberg, John Wiley, New York, 1997, p. 116.

112 *with the headline:* "A Breakthrough Drug for Depression," by Geoffrey Cowley, *Newsweek,* March 26, 1990.

112 *the inevitable backlash:* For a detailed account of the negative publicity concerning Prozac and the subsequent FDA hearings, see *Talking Back to Prozac,* by Peter Breggin and Ginger Breggin, St. Martin's Press, New York, 1994.

112 *Prozac was second:* "Drugs Sales Can Leave Elderly a Grim Choice: Pills or Other Needs," by Lucette Lagnado, *Wall Street Journal,* November 17, 1998, p. A15.

112 *Worldwide:* Eli Lilly public relations provided me with this information on worldwide sales by telephone on August 12, 1998.

112 *The market for SSRIs is growing:* "Blooming Business for Happy Pills," by Susan Aldridge, *Chemistry and Industry,* December 1, 1997.

113 *children age twelve and younger:* "Use of Antidepression Medicine for Young Patients Has Soared," by Barbara Strauch, *New York Times,* August 10, 1997, p. A1. According to the article, sales of Prozac to children six to twelve years old rose 298 percent from 1995 to 1996.

113 *there "has never been":* "Efficacy of Antidepressant Medication with Depressed Youth: What Psychologists Should Know," by John Sommers-Flanagan and Rita Sommers-Flanagan, *Professional Psychology: Research and Practice* 27, no. 2, 1996, pp. 145–153.

113 *peppermint-flavored version:* "Peppermint Prozac," by Arianna Huffington, *U.S. News and World Report,* August 18, 1997, p. 28.

113 *"cosmetic psychopharmacology":* "The New You," by Peter Kramer, *Psychiatric Times,* March 1990, pp. 45–46.

113 *for twenty-one weeks:* Breggin and Breggin, *Talking Back to Prozac,* p. 3.

113 *I watched him participate:* The symposium at the New School took place on October 5, 1995.

114 *"She looked different":* Kramer, *Listening to Prozac,* p. 7. As an antidote to the cheerful case histories in *Listening to Prozac,* see "Anatomy of Melancholy," by Andrew Solomon, *New Yorker,* January 19, 1998, pp. 46–61. The author, a novelist, recounted his harrowing struggle with depression.

115 *Kramer cautioned:* Kramer, *Listening to Prozac,* p. 11.

116 *a psychologist in Wenatchee:* This incident is recounted in Breggin and Breggin, *Talking Back to Prozac,* p. 6–7.

116 *Prozac could widen the gap:* "The Coverage of Happiness," by Robert Wright, *New Republic,* March 14, 1994, pp. 24–29.

116 *the results of a comparison:* "Initial Antidepressant Choice in Primary Care," by Gregory Simon et al., *Journal of the American Medical Association,* June 26, 1996, 1897–1902.

116 *forty-two separate trials:* "Selective Serotonin Reuptake Inhibitors: Meta-Analysis of Discontinuation Rates," by S. A. Montgomery et al., *International Clinical Psychopharmacology* 9, 1995, pp. 47–53.

116 *"The overall number":* "Are the SSRIs Really Better Tolerated Than the TCAs for Treatment of Depression?" by J. C. Nelson, *Psychiatric Annals* 24, 1994, p. 631, cited in Fisher and Greenberg, *From Placebo to Panacea,* p. 124.

116 *Early reports on the SSRIs:* Findings on the sexual side effects of Prozac and other drugs were described in "When Depression Lifts But Sex Suffers," by Jane Brody, *New York Times,* May 15, 1996, p. C7. See also "Sex and the Depressed Patient," by Robert Segraves, *Current Canadian Psychiatry and Neurology,* May 1995, pp. 7–13.

117 *premature ejaculation: Sexual Pharmacology: Drugs That Affect Sexual Function,* by Theresa Crenshaw and James Goldberg, Norton, New York, 1996, p. 286.

117 *the opposite effect on bivalves:* "Prozac Works on Clams and Mussels," *Science News,* January 24, 1998, p. 63.

117 *the fine print, literally:* Kramer, *Listening to Prozac,* p. 366.

117 *psychotherapy and drugs produce:* "Psychotherapy Versus Medication for Depression: Challenging the Conventional Wisdom with Data," by David Antonuccio et al., *Professional Psychology: Research and Practice* 26, no. 6, 1995, pp. 574–585.

118 New York Times *announced:* "Psychotherapy Is As Good As Drug in Curing Depression, Study Finds," by Philip Boffey, *New York Times,* May 14, 1986, p. A1.

119 *drugs be the initial treatment:* "Science Is Not a Trial (But It Can Sometimes Be a Tribulation)," by Irene Elkin et al., *Journal of Consulting and Clinical Psychology* 64, no. 4, 1996, p. 92.

119 *"Although there was":* "The NIMH Treatment of Depression Collaborative Research Program: Where We Began and Where We Are," by Irene Elkin, *Handbook of Psychotherapy and Behavior Change,* 4th ed., edited by A. D. Bergin and S. L. Garfield, John Wiley, New York, 1994, p. 130.

119 *"there was no indication":* Ibid., p. 125.

119 *One "striking" finding:* Ibid., p. 131.

119 *Kramer . . . once told me:* I interviewed Kramer by telephone in July 1996.

120 *Harry Stack Sullivan forced:* Shorter, *A History of Psychiatry,* p. 205.

120 *survey carried out in 1995:* "Mental Health: Does Therapy Help?" *Consumer Reports,* November 1995, pp. 734–739.

120 *declared Martin Seligman:* "The Effectiveness of Psychotherapy," by Martin Seligman, *American Psychologist,* December 1995, pp. 965–974.

121 *therapeutic power of religious belief:* "Religiosity and Remission of Depression in Medically Ill Older Patients," by Harold Koenig et al., *American Journal of Psychiatry,* April 1998, pp. 536–542.

121 *Roger Greenberg and Seymour Fisher:* Fisher and Greenberg discuss antidepressant research at length in their chapter, "Mood-Mending Medicines: Probing Drug, Psychotherapy and Placebo Solutions," and other sections of *From Placebo to Panacea.*

122 *patient ratings alone:* "A Meta-analysis of Antidepressant Outcome Under 'Blinder' Conditions," by Roger Greenberg et al., *Journal of Consulting and Clinical Psychology* 60, no. 5, 1992, pp. 664–669.

122 *"the conventional claims":* Fisher and Greenberg, *From Placebo to Panacea,* p. 362. The conclusions of Greenberg and Fisher are corroborated by "Listening to Prozac But Hearing Placebo," by Irving Kirsch and Guy Sapirstein, *Prevention and Treatment* 1, article 0002a. The American Psychological Association has made this article available on the World Wide Web at http://journals.apa.org/prevention/volume1/pre0010002a.html.

122 *Early reports on lithium:* Fisher and Greenberg, *From Placebo to Panacea,* p. 157.

122 *but a 1990 review:* Ibid., p. 150.

122 *Jamison's memoir: An Unquiet Mind,* by Kay Jamison, Vintage Books, New York, 1995.

123 *some of the trials:* Fisher and Greenberg, *From Placebo to Panacea,* p. 150.

123 *"Unfortunately, after scrutinizing":* "A Re-examination of the Placebo-Controlled Trials of Lithium Prophylaxis in Manic-Depressive Disorder," by J. Moncrieff, *British Journal of Psychiatry* 167, 1995, p. 572, cited in Fisher and Greenberg, *From Placebo to Panacea,* p. 156.

123 *"The history of the research":* Ibid., p. 157.

123 *a leading psychiatric textbook:* From "Chlorpromazine Is No Penicillin," by Scull, p. 9.

124 *as many as 40 percent: Medications for the Treatment of Schizophrenia: Questions and Answers,* DHHS Publication No. (ADM) 92–1950, edited by Deborah Dauphinais, National Institute of Mental Health, Bethesda, Maryland, 1992, p. 5. I am indebted to Walter Brown of Brown University for helping me understand the difference between extrapyramidal effects and tardive dyskinesia.

124 *Claims that newer neuroleptics:* See "A Critique of the Use of Neuroleptic Drugs in Psychiatry," by David Cohen, in Fisher and Greenberg, *From Placebo to Panacea,* pp. 173–228.

124 *the anti-psychotic medication risperidone:* Ibid., p. 176.

125   *"it is too early to tell"*: Ibid., p. 213.

125   *"Are we suggesting"*: Ibid., p. 371.

125   *Brown acknowledged:* "The Placebo Effect," by Walter Brown, *Scientific American,* January 1998, pp. 93.

126   *a startling proposal:* "Placebo as a Treatment for Depression," by Walter Brown, *Neuropsychopharmacology* 10, no. 4, 1994, pp. 265–288. Brown's article was followed by responses from readers and a final response from Brown.

126   *a 1965 study:* Ibid., p. 267. The study Brown cited was "Nonblind Placebo Trial: An Exploration of Neurotic Patients' Responses to Placebo When Its Inert Content Is Disclosed," by L. C. Park and L. Covi, *Archives of General Psychiatry* 12, 1965, pp. 336–345.

127   *"Mrs. Jones":* Brown, "Placebo as a Treatment for Depression," p. 267.

127   *"would play into":* Ibid., p. 272.

127   *Another psychiatrist wondered:* Ibid., p. 280.

127   *"A thread that runs through alternative medicine":* Ibid., p. 288.

128   *In his 1998 book: Last Resort,* by Jack Pressman, Cambridge University Press, New York, 1998.

128   *reported in* Neurosurgery: Cited in "Lobotomy's Back," by Frank Vertosick, *Discover,* October 1997, pp. 66–72. See also "Psychosurgery Redux," by Wray Herbert, *U.S. News and World Report,* November 3, 1997, pp. 63–64. The Cingulotomy Unit at Massachusetts General Hospital describes its work on a Web site, http://brain.mgh.harvard.edu:100/cingulot.htm.

129   *Sackheim is a slim:* I interviewed Harold Sackheim at the New York State Psychiatric Institute on October 1, 1997. Sackheim and two coauthors provided an excellent review of shock therapy in "Electroconvulsive Therapy," in *Psychopharmacology: The Fourth Generation of Progress,* edited by Floyd Bloom and David Kupfer, Raven Press, New York, 1995, Chapter 95.

129   *"Not a single controlled study":* Shorter, *A History of Psychiatry,* p. 285.

130   *"people that you see on TV":* One show business personality who has publicly extolled the benefits of electroconvulsive therapy for treating his own depression is Dick Cavett. In an interview published in *People* on August 3, 1992, the former talk-show host called shock therapy "miraculous," "like a magic wand."

130   *patients' unconscious desire:* See Sackheim et al., "Electroconvulsive Therapy," p. 1134.

131   *to observe patients:* I observed patients being treated with electroconvulsive therapy at the New York State Psychiatric Institute on October 10, 1997.

134   *success rates ranging:* "Setting the ECT Stimulus," *Psychiatric Times,* p. 1, June 1995, p. 1.

134   *psychiatric drugs to children:* See *The War Against Children,* by Peter

Breggin and Ginger Ross Breggin, St. Martin's Press, New York, 1994.

136 *"new biological materialism"*: Kramer, *Listening to Prozac*, p. xiv.

Chapter 5: GENE-WHIZ SCIENCE

137 *"Oedipus Schmoedipus"*: "Oedipus, Schmoedipus. The Fault, Dear Sigmund, May Be in Our Genes," by James Collins, *Time*, December 9, 1996, p. 74.

138 *Ashkenazi Jews:* Information on Tay-Sachs disease can be found at the Web site of the March of Dimes Foundation: http://www.noah.cuny.edu/pregnancy/march_of_dimes/birth_defects/taysachs.html.

139 *talk show* Donahue: "How to Tell If Your Child's a Serial Killer," *Donahue*, aired on February 25, 1993.

139 *a report on violence: Understanding and Preventing Violence*, edited by Albert Reiss and Jeffrey Roth, National Academy Press, Washington, D.C., 1993. See also the history of double-Y syndrome in *The Mismeasure of Man*, by Stephen Jay Gould, Norton, New York, 1981, pp. 143–145.

140 *"the ultimate answer"*: Quoted in *Forbidden Knowledge*, by Roger Shattuck, Harcourt Brace, New York, 1996, p. 178.

140 *"We used to think"*: "Happy Birthday, Double Helix," by Leon Jaroff, *Time*, March 15, 1993, p. 57.

140 *The biologist Daniel Koshland:* Daniel Koshland's editorials on behavioral genetics in *Science* included "Nature, Nurture, and Behavior," March 20, 1987; "Sequences and Consequences of the Human Genome," October 13, 1989; and "The Rational Approach to the Irrational," October 12, 1990.

140 *a disturbing rightward drift:* See, for example, *The DNA Mystique*, by Dorothy Nelkin and M. Susan Lindee, W. H. Freeman, New York, 1995.

141 *Psychoanalysis "would not"*: "What, Me Not Worry?" by Adam Phillips, *New York Times*, December 13, 1996, p. A39.

142 *One of the earliest:* For an excellent history of eugenics, including information about Galton, Davenport, and other early enthusiasts, see *In the Name of Eugenics*, by Daniel Kevles, Knopf, New York, 1985.

142 *continued forcibly sterilizing:* See "Here, of All Places," *Economist*, August 30, 1997, p. 36.

143 *identical-twin studies:* For an excellent account of the twin studies done at the University of Minnesota and elsewhere, see *Twins*, by Lawrence Wright, John Wiley, New York, 1997.

144 *"On multiple measures"*: "Sources of Human Psychological Differences: The Minnesota Study of Twins Reared Apart," by Thomas Bouchard et al., *Science*, October 12, 1990, p. 223.

144 *This message was spelled out:* Wright, *Twins*, p. 143. What made Wright's sweeping conclusion especially puzzling was that his book presented a great deal of evidence that undermined such hard-core genetic determinism.

145 *the psychologist Susan Farber:* Wright, *Twins*, pp. 69–70.

145 *Other critics have contended:* For critiques of the Minnesota twin research, see "The Genetic Analysis of Human Behavior: A New Era?" by Paul Billings et al., *Social Science and Medicine* 35, no. 3, 1992, pp. 227–238; and "Bewitching Science," by Val Dusek, *Science for the People*, November–December 1987, pp. 19–22.

145 *Leon Kamin:* I interviewed Leon Kamin several times by telephone and in person in 1993.

145 *Whereas some news accounts:* Stories that reported erroneously that the two British giggle sisters and the Nazi and Jewish brothers met for the first time in Minneapolis included "Identical Twins Reared Apart," by Constance Holden, *Science*, March 21, 1980, pp. 1323–1328; and "The Twins," by Cynthia Gorney, *Washington Post*, December 10, 1979. Leon Kamin contended, and I confirmed through telephone interviews, that the giggle sisters and the Nazi and Jewish brothers had met and corresponded before being brought to Minnesota. For more on the role of publicity in the Minnesota research, see "Scientists Split over Twins as Lab Subjects," by Gail Golden, *Chicago Tribune*, September 6, 1988, p. C1.

146 *sold their life story:* I was told by a Los Angeles producer named Anthony Mason in 1993 that he had bought an option to the rights to the story of Oskar and Jack and had resold it to Hearst Entertainment.

146 *the Pioneer Fund, a remnant:* See Wright, *Twins*, p. 57.

146 *Lykken complained:* Ibid., pp. 131–132.

146 *"Now, why would anyone":* "What Did They Name the Dog?" by Wendy Doniger, *London Review of Books*, March 19, 1998, p. 32.

147 *Schizophrenia is typical:* Statistics about the heritability of schizophrenia are drawn from "Genetic Basis of Schizophrenia," by Peter McGuffin et al., *Lancet*, September 9, 1995, pp. 678–682.

147 *Nevertheless, encouraged by the successful assaults:* For an excellent summary of the methods used for identifying single genes, see *Genome*, by Jerry Bishop and Michael Waldholz, Simon & Schuster, New York, 1990.

148 *a gene in chromosome 11:* "Bipolar Affective Disorders Linked to DNA Markers on Chromosome 11," by Janice Egeland et al., *Nature*, February 26, 1987, pp. 783–787.

148 *manic depression in three Israeli families:* "Genetic Linkage Between X-Chromosome Markers and Bipolar Affective Illness," by Miron Baron et al., *Nature* 326, 1993, pp. 289–292.

148 *A more extensive analysis of the Amish families:* "Re-evaluation of the

Linkage Relationship Between Chromosome 11p Loci and the Gene for Bipolar Affective Disorder in the Old Order Amish," by John Kelsoe et al., *Nature*, November 16, 1989, pp. 238–243.

148   *the results from the Israeli families:* "Diminished Support for Linkage Between Manic Depressive Illness and X-Chromosome Markers in Three Israeli Pedigrees," by Miron Baron et al., *Nature Genetics*, January 1993, pp. 49–55.

148   *Studies of schizophrenia:* The two contradictory papers on schizophrenia, both published in *Nature*, November 10, 1988, were "Localization of a Susceptibility Locus for Schizophrenia on Chromosome 5," by R. Sherrington et al., pp. 164–167; and "Evidence Against Linkage of Schizophrenia to Markers on Chromosome 5 in a Northern Swedish Pedigree," by James Kennedy et al., pp. 167–170.

148   *British group quietly retracted:* Hugh Gurling of University College and Middlesex School of Medicine, London, an author of the 1988 *Nature* paper linking schizophrenia to chromosome 5, told me in a telephone interview in 1993 that new data no longer supported the linkage.

148   *Peter McGuffin:* I interviewed Peter McGuffin by telephone on March 3, 1998.

149   *a working group:* "Panel Urges Caution on Genetic Testing for Mental Disorders," by David Dickson, *Nature*, September 24, 1998, p. 309.

150   *Evidence "is rapidly accumulating":* **Freudian Fraud**, by E. Fuller Torrey, HarperCollins, New York, 1992, p. 227.

150   *"When I laid stress":* **The Freud Reader**, edited by Peter Gay, Norton, New York, 1989, p. 37.

150   *Torrey has proposed:* Torrey's ideas about schizophrenia are described in "Schizophrenia's Most Zealous Foe," by Michael Winerip, *New York Times Magazine*, February 22, 1998, pp. 26–29.

150   *German virologists:* See "It Kills Horses, Doesn't It?" by Robert Kunzig, *Discover*, October 1997, pp. 97–105. (Mindy Kornhaber of Harvard University, upon reading this material, suggested that cows' neurological conditions be called "moo-ed disorders.")

151   *prions to schizophrenia:* "A Prion-Based Psychiatric Disorder," by Helena Samaia et al., *Nature*, November 20, 1997, p. 241.

151   *prenatal traumas:* "New Culprits Cited for Schizophrenia," by Bruce Bower, *Science News*, February 3, 1996, p. 68.

151   *group led by Kenneth Blum:* "Allelic Association of Human Dopamine D2 Receptor Gene in Alcoholism," by Kenneth Blum et al., *Journal of the American Medical Association*, April 18, 1990, pp. 2055–2060.

151   *a front-page article:* "Scientists See a Link Between Alcoholism and a Specific Gene," by Lawrence Altman, *New York Times*, April 18, 1990, p. A1.

152   *"no physiologically significant":* "The A1 Allele at the D2 Dopamine Receptor Gene and Acoholism: A Reappraisal," by Joel Gelernter,

David Goldman, and Neil Risch, *Journal of the American Medical Association*, 269, 1993, pp. 1673–1677.

152   *One of the authors:* I spoke to Neil Risch, then at Yale University, by telephone in 1993.

152   *Irving Gottesman:* I interviewed Gottesman by telephone in 1993.

152   *the D2 marker:* For a list of conditions linked to the D2 marker, see "Reward Deficiency Syndrome," by Kenneth Blum et al., *American Scientist*, March–April 1996, pp. 132–145.

152   *Hamer held out the possibility: Living with Our Genes,* by Dean Hamer and Peter Copeland, Doubleday, New York, 1998, p. 144.

152   *In 1993, Hamer:* "A Linkage Between DNA Markers on the X Chromosome and Male Sexual Orientation," by Dean Hamer et al., *Science* 261, 1993, pp. 321–327.

153   *to cowrite a book: The Science of Desire,* by Dean Hamer and Peter Copeland, Simon & Schuster, New York, 1994.

153   *replicated the X-chromosome result:* "Linkage Between Sexual Orientation and Chromosome Xq28 in Males But Not in Females," by S. Hu et al., *Nature Genetics* 11, 1995, pp. 248–256.

153   *In 1995 George Ebers:* The findings of Ebers and Rice were first reported in "NIH's 'Gay Gene' Study Questioned," by Eliot Marshall, *Science,* June 30, 1995, p. 1841. The article also revealed that Hamer was being investigated by the Office of Research Integrity of the Department of Health and Human Services. The investigation of Hamer was first made public in "Study on 'Gay Gene' Challenged," by John Crewdson, *Chicago Tribune,* June 25, 1995, p. C1. Crewdson reported that a coauthor of Hamer's 1993 paper on male homosexuality had accused him of improper handling of data. The investigation was later dropped without charges being filed against Hamer.

153   *study of fifty-four pairs:* "Genetic Linkage Study of Male Homosexual Orientation," by Alan Sanders et al., was presented as a poster at the annual meeting of the American Psychiatric Association in Toronto, June 1998. Sanders moved from the National Institute of Mental Health to the University of Chicago in early 1999.

153   *Hamer reinterpreted:* Hamer, *Living with Our Genes,* p. 197. The paper on novelty seeking that Hamer coauthored was "Population and Familial Association Between the D4 Dopamine Receptor Gene and Measures of Novelty Seeking," by J. Benjamin et al., *Nature Genetics* 12, 1996, pp. 81–84. The same issue included another article making a similar claim: "Dopamine D4 Receptor (D4DR) Exon III Polymorphism Associated with the Human Personality Trait of Novelty Seeking," by R. Ebstein et al., pp. 78–80. The paper on anxiety that Hamer coauthored was "Association of Anxiety-Related Traits with a Polymorphism in the Serotonin Transporter Gene-

Regulatory Region," by Klaus-Peter Lesch et al., *Science* 274, 1996, pp. 1527–1531.

154 *no evidence for the novelty-seeking gene:* See "The Association Between the Dopamine D4 Receptor (D4DR) 16 Amino Acid Repeat Polymorphism and Novelty Seeking," by Anil Malhotra et al., *Molecular Psychiatry* 1, 1996, pp. 388–391; and "Human Novelty-Seeking Personality Traits and Dopamine D4 Receptor Polymorphisms: A Twin and Genetic Association Study," by Michael Pogue-Geile et al., *American Journal of Medical Genetics* 81, 1998, pp. 44–48. The Pogue-Geile study was first reported on in "Born Happy?" by Sharon Begley, *Newsweek*, October 14, 1996, p. 79.

154 *Cyril Burt:* An account of the Cyril Burt affair can be found in Kevles, *In the Name of Eugenics*.

154 *Bernie Devlin:* "The Heritability of IQ," by Bernie Devlin et al., *Nature*, July 31, 1997, pp. 468–471. See also "Wombs with a View," by Sharon Begley, *Newseek*, August 11, 1997, p. 61.

154 *Burt's contention: The Bell Curve*, by Charles Murray and Richard Herrnstein, Free Press, New York, 1994.

155 *Critics raised numerous:* For critiques of *The Bell Curve*, see "Curveball," by Stephen Jay Gould, *New Yorker*, November 28, 1994, pp. 139–149; and the special issue of *The New Republic*, October 31, 1994.

156 *"Surely people differ": Language and Problems of Knowledge*, by Noam Chomsky, MIT Press, Cambridge, 1988, p. 164.

156 *Flynn effect:* I first found a discussion of the Flynn effect in "Intelligence: Knowns and Unknowns," a report of the American Psychological Association released in August 1995. Later I learned that Murray and Herrnstein had briefly mentioned the Flynn effect (and casually dismissed it as irrelevant to their argument) in *The Bell Curve*. Flynn presented his data in "Massive IQ Gains in 14 Nations: What IQ Tests Really Measure," *Psychological Bulletin* 101, 1987, pp. 171–191. See also my story, "Get Smart, Take a Test," *Scientific American*, November 1995, pp. 12–13. I interviewed Flynn by e-mail and fax in 1995.

158 *The psychologist Arthur Jensen:* I interviewed Arthur Jensen by telephone in 1995.

159 *the fifteen-point gap … may be closable:* See "The Black-White Test Score Gap: Why It Must Be Closed. Why It Can Be," by Christopher Jencks and Meredith Phillips, *The American Prospect*, September–October 1998, pp. 44–53.

160 *"geniuses are born":* "Genning Up on Genius Genes," by Robert Matthews, *Sunday Telegraph*, January 24, 1993, p. 9. The story reported on a conference in London at which Plomin had presented his results.

160 *Psychological Science:* "A Quantitative Trait Locus Associated with

Cognitive Ability in Children," by Robert Plomin et al., *Psychological Science* 9, 1998, pp. 159–166. See also "The Genetics of Cognitive Abilities and Disabilities," by Robert Plomin and John Defries, *Scientific American*, May 1998, pp. 62–69.

160 *"I confidently predict":* The quotation is from John Kihlstrom, a psychologist at the University of California at Berkeley, in "First Gene to Be Linked with High Intelligence Is Reported Found," by Nicholas Wade, *New York Times*, May 14, 1998.

161 *"Whether anyone thinks":* Hamer, *Living with Our Genes*, p. 301.

161 *"'Intelligence' does not do justice":* Remaking Eden, by Lee Silver, Avon Books, New York, 1997, p. 249.

162 *"Except for anecdotal reports":* "Human Gene Therapy," by W. French Anderson, *Nature*, supplement to vol. 392, April 30, 1998, p. 25.

162 *Robert Weinberg:* Robert Weinberg was quoted in "Hype Surrounds Genomics Inc.," *Science*, February 7, 1997, p. 770.

162 *One of the wiser investigators:* I interviewed Jerome Kagan at Harvard University on November 18, 1997.

164 *"I do not believe":* Galen's Prophecy, by Jerome Kagan, Basic Books, New York, 1994, p. xxi.

164 *"who consistently protected":* Ibid., p. 263.

165 *"incessantly struck":* The quotation from Darwin is in *In Search of Human Nature*, by Carl Degler, Oxford University Press, New York, 1991, p. 15. Degler's book is a treasure trove of information about the history of genetic determinism in science and society.

166 *"To aid the bad":* Ibid., p. 11.

166 *man "must remain subject":* Darwin's letter was reprinted in "A Recently Discovered Darwin Letter on Social Darwinism," by Richard Weikart, *Isis* 86, 1995, pp. 609–611.

166 *"provides a solution":* Why Freud Was Wrong, by Richard Webster, Basic Books, 1995, p. 457.

PAGE **Chapter 6: DARWIN TO THE RESCUE!**

167 *"But then arises the doubt":* The quotation from Darwin is in *Bright Air, Brilliant Fire*, by Gerald Edelman, Basic Books, New York, 1992, p. 42.

167 *No, this was the annual meeting:* The Human Behavior and Evolution Society met at the University of California at Santa Barbara June 28–July 2, 1995.

167 *Charles Darwin's prophecy: On the Origin of Species*, by Charles Darwin, Harvard University Press, 1964, p. 488.

169 *anthropologist Lee Cronk:* In his lecture, titled "The Bathwater and the Baby: What the Culture Concept Can and Cannot Do for Human Behavioral Ecology," Cronk singled out the anthropologist Clifford Geertz of the Institute for Advanced Study for ridicule.

169   *"stone agers in the fast lane":* The phrase was coined by Boyd Eaton of Emory University.

170   *waist-to-hip ratio of 0.7:* Singh first presented his findings in "Adaptive Significance of Female Physical Attractiveness," *Journal of Personality and Social Psychology* 65, 1993, pp. 293–307. Data that undermined Singh's thesis were reported in "Is Beauty in the Eye of the Beholder?" by Douglas Yu and Glenn Shepard, *Nature*, November 26, 1998, pp. 321–322.

170   *Yes, we do!* I interviewed Cosmides and Tooby by telephone in May 1995. I also spoke to them at the HBES meeting in June and communicated with them by fax.

170   *"social sciences are still adrift":* The Adapted Mind, edited by Jerome Barkow, Leda Cosmides, and John Tooby, Oxford University Press, New York, 1992, p. 23.

173   *Critics of evolutionary psychology:* One of the most visible critics of evolutionary psychology is Stephen Jay Gould of Harvard University. See his op-ed essay "Let's Leave Darwin Out of It," *New York Times*, May 29, 1998. Gould also attacked evolutionary psychology in two consecutive articles in the *New York Review of Books*, "Darwinian Fundamentalism," June 12, 1997, and "Evolution: The Pleasures of Pluralism," June 26, 1997. See also the exchange of letters published in the August 14 and October 9 issues. Gould granted that humans "are animals, and the mind evolved, therefore, all curious people must support the quest for an evolutionary psychology. But the movement that has commandeered this name adopts a fatally restrictive view of the meaning and range of evolutionary explanations." If evolutionary psychologists keep pressing their most extravagant claims, Gould predicted, "they will eventually suffer the fate of the Freudians, who also had some good insights but failed spectacularly, and with serious harm imposed upon millions of people (women, for example, who were labeled as 'frigid' when they couldn't make an impossible physiological transition from clitoral to vaginal orgasm), because they elevated a limited guide into a rigid creed that became more of an untestable and unchangeable religion than a science."

    Unfortunately, Gould's polemic—and the responses to it—generated more smoke than illumination. He spent less energy on analyzing evolutionary psychology than on settling old scores and promoting his own highly derivative contributions to evolutionary theory. These include punctuated equilibrium, which posits that evolution proceeds in fits and starts; contingency, the notion that evolution is shaped not only by natural selection but also by asteroid impacts and other unpredictable, random events; and spandrels, which are nonadaptive by-products of evolution. (*Spandrel* was originally an architectural term referring to the triangular space between an arch and its surrounding structure.) Gould's rhetoric was also distracting. At one point he mocked two of his opponents (the journalist Robert Wright and the philosopher Daniel Dennett): "Right after King Henry's stirring Saint Crispin's Day speech on the battlefield of Agincourt, Shakespeare supplies some humorous relief, as Falstaff's former servant Pistol extracts a ransom by loud bluff and posturing. Pistol's own servant then makes the famous observation: 'The saying is true—the empty vessel makes the greatest sound.'" Gould wrapped up his diatribe with a similar riposte: "But as T. H. Huxley said of Richard Owen, in a parody of Dryden's line about Alexander the Great refighting all his battles during a drunken monologue—'And thrice he routed all his foes, and thrice he slew the slain'—life is just too short for occupying oneself with the slaying of the slain more than twice." Gould apparently intended to impress readers of the *New York Review* with his erudition, but he seemed to be risking—indeed, committing—self-parody.

174   *I first saw Pinker: The Language Instinct*, by Steven Pinker, Harper Perennial, New York, 1994. I interviewed Pinker in Cambridge, Massachusetts, on August 6, 1997.

175   *"In rummaging through": How the Mind Works*, by Steven Pinker, Norton, New York, 1997, p. 22.

175   *the size of testicles:* Ibid., p. 465.

175   *"I hate to lose":* Ibid., p. 392.

176   *even the Dalai Lama:* Ibid., p. 519.

176   *"Women really look":* Ibid., p. 482.

176   *"bad evolutionary 'explanations'":* Ibid., pp. 37–38.

176   *"an exquisite confection":* Ibid., p. 534.

176   *"a desperate measure":* Ibid., p. 556.

176   *"Fictional narratives supply us":* Ibid., p. 543.

176   *article about teenage girls:* "The Thin Red Line," by Jennifer Egan, *New York Times Magazine,* July 27, 1997.

178   *Chomsky insisted:* I discussed evolutionary psychology with Noam Chomsky by telephone in 1995. Chomsky discussed evolutionary theory and human behavior in *Language and Problems of Knowledge,* by Noam Chomsky, MIT Press, Cambridge, 1988. See also my treatment of Chomsky in *The End of Science,* Broadway Books, New York, 1996, pp. 149–154.

180   *"raised too many hackles":* See "'Sociobiology' to History's Dustbin?" *Science,* July 19, 1996, p. 315. Randolph Nesse of the University of Michigan, a founder of the HBES, also told me in 1995 that he and other founders deliberately rejected the term *sociobiology* because of its negative connotations.

180   *according to Cosmides and Tooby:* Cosmides and Tooby discussed the differences between evolutionary psychology and sociobiology in

a fax they sent to me in 1995. See also the section titled "Why I Am Not a Sociobiologist" in "Evolutionary Psychology: A New Paradigm for Psychological Science," by David Buss, *Psychological Inquiry*, 1995, pp. 1–30.

180 *"I don't have anything":* Richard Alexander made these remarks to me in a telephone interview in 1995.

181 *group selection fell out of fashion:* See *Adaptation and Natural Selection: A Critique of Some Current Evolutionary Thought,* by George Williams, Princeton University Press, Princeton, New Jersey, 1996. Lately, several evolutionary theorists have tried to resurrect group selection. See *Unto Others,* by Elliott Sober and David Sloan Wilson, Harvard University Press, Cambridge, 1998.

181 *the theory of kin selection:* Hamilton presented his theory of kin selection and altruism in "The Evolution of Altruistic Behavior," *American Naturalist* 97, 1963, pp. 354–356.

181 *"two brothers or eight cousins":* The quotation is in Pinker, *How the Mind Works,* p. 400.

181 *reciprocal altruism:* "The Evolution of Reciprocal Altruism," by Robert Trivers, *Quarterly Review of Biology* 46, 1971, pp. 35–57.

181 *"Why so many churches, priests, judges":* "The Softer Side of Sociobiology," by H. Allen Orr, *Boston Review of Books,* October–November 1997, pp. 44.

182 *Cosmides has argued:* "The Logic of Social Exchange: Has Natural Selection Shaped How Humans Reason?" by Leda Cosmides, *Cognition* 31, 1989, pp. 187–276.

182 *Fetzer, a philosopher:* I interviewed James Fetzer at the 1995 HBES meeting in Santa Barbara.

182 *Mithen, a British anthropologist:* Steven Mithen presented his objections to evolutionary psychology in a lecture at the HBES meeting. He expanded on these ideas in *The Prehistory of the Mind,* Thames and Hudson, London, 1996. For another critique of evolutionary psychology, see "The Adaptive Nature of the Human Neurocognitive Architecture," by Peggy La Cerra and Roger Bingham, *Proceedings of the National Academy of Sciences* 95, 1998, pp. 11290–11294.

183 *females are instinctively more coy: The Evolution of Desire,* by David Buss, Basic Books, New York, 1994. My critique of Buss's ideas about male and female sexuality is based on comments made to me by the anthropologist Elizabeth Blaffer Hrdy of the University of California in 1995.

183 *the murder of children:* See *Homicide,* by Martin Daly and Margo Wilson, Aldine de Gruyter, Hawthorne, New York, 1988; and "Evolutionary Social Psychology and Family Homicide," by Martin Daly and Margo Wilson, *Science,* October 28, 1988, pp. 519–524.

184 *from mice to monkeys:* See "Evolutionists Take the Long View on Sex and Violence," *Science,* August 20, 1993, p. 987.

184   *Even Wilson and Daly have warned:* I interviewed Wilson and Daly several times by telephone in 1995.

184   *Pinker singled out* Homicide: "Tales Twice, Indeed Thrice," *New York Times,* December 6, 1997, p. B9.

184   *Pinker later wrote:* "Why They Kill Their Newborns," by Steven Pinker, *New York Times Magazine,* November 2, 1997, pp. 52–54. The letter from Claude Fischer responding to Pinker's article was published on November 23, 1997.

185   *"Heritable behavioral variation is":* The quotation is from an unpublished paper, "Evolutionary Adaptationism: Another Biological Approach to Criminal and Antisocial Behavior," sent to me in 1996 by Martin Daly.

186   *But evolutionary psychologists:* For a discussion of evolutionary psychology versus behavioral genetics, see the comments by David Lykken, a behavioral geneticist at the University of Minnesota, on the Edge Web site, www.edge.org. Lykken complained that evolutionary psychologists make the "odd mistake" of "arbitrarily assuming that all of the genetic diversity that permitted natural selection to evolve the human brain has now been exhausted and that individual psychological differences that we observe today are all environmentally produced. . . . In other words, all human babies today, unlike Paleolithic times, have brains like so many new Mac computers, waiting to be programmed. This is wildly improbable on evolutionary grounds. E.g., if there is abundant genetic diversity in the psychology of domestic animals, as Darwin himself observed, as well as in the anatomy and physiology of humans, as any child can observe, why is the human brain the sole exception?"

187   *might "increase creativity": Why We Get Sick: The New Science of Darwinian Medicine,* by Randolph Nesse and George Williams, Times Books, New York, 1994, p. 225.

187   Evolutionary Psychiatry: *Evolutionary Psychiatry: A New Beginning,* by Anthony Stevens and John Price, Routledge, London, 1996, was reviewed in "Darwin on the Brain," by Steven Rose, *Nature,* April 3, 1997, pp. 454–455.

187   *anthropologist Napoleon Chagnon:* Chagnon presented his findings on the link between male violence and number of offspring in "Life Histories, Blood Revenge, and Warfare in a Tribal Population," *Science,* February 26, 1988, pp. 985–992. See also his classic work *Yanomamo: The Fierce People,* Holt, Rinehart and Winston, New York, 1968. Chagnon and I discussed his similarity to Stephen Jay Gould in a telephone interview in 1995.

189   *According to Sulloway: Born to Rebel,* by Frank Sulloway, Pantheon Books, New York, 1996. I first heard Sulloway present his birth-order theory at the HBES meeting in Santa Barbara on June 30, 1995. Positive reviews of *Born to Rebel* included "The Birth of an Idea," by

Robert Boynton, *New Yorker*, October 7, 1996, p. 72; and "First Born, Later Born," by Geoffrey Cowley, *Newsweek*, October 7, 1996, pp. 68–74. Critical reviews included "Family Niche and Intellectual Bent," by John Modell, *Science*, January 31, 1997, pp. 624–625; and "Birth Order, Schmirth Order," by Alan Wolfe, *New Republic*, December 23, 1996, pp. 29–35.

191 *Even Stephen Jay Gould:* Sulloway claimed in an interview on the Internet site the Edge (www.edge.org) that Gould had spoken favorably about the birth-order thesis on the television program *Nightline.*

191 *1983 book: Birth Order: Its Influence on Personality*, by Cecile Ernst and Jules Angst, Springer-Verlag, Berlin, Germany, 1983.

191 *"a sheer waste of time and money":* The quotation from Ernst and Angst is found in comments by Judith Harris on the Edge Web site, www.edge.org.

192 *"Birth order effects are frequently found": The Nurture Assumption*, by Judith Harris, Free Press, New York, 1998, pp. 375. See also the exchange between Sulloway and Harris on the Edge, www.edge.org. I found Harris more persuasive when attacking Sulloway than when defending the main theme of her book: that children's personalities are shaped primarily by their genes and their peers rather than by their parents.

192 *Pinker told me:* Steven Pinker expressed his doubts about *Born to Rebel* when I interviewed him in Cambridge on August 6, 1997.

193 *1979 book: Freud, Biologist of the Mind*, by Frank Sulloway, Basic Books, New York, 1979.

193 *Evolutionary psychologists have more in common:* For a fascinating comparison of evolutionary psychology and psychoanalysis, see *The Moral Animal*, by Robert Wright, Pantheon Books, New York, 1994, pp. 313–326. The British author Christopher Badcock attempted (not very persuasively) to reconcile psychoanalysis and evolutionary theory in *Oedipus in Evolution*, Basil Blackwell, Oxford, 1990. For an amusing review of Badcock's book, see "Translation from the Greek," by V. Reynolds, *Nature*, May 24, 1990, p. 301.

193 *The most effective liars:* See *Social Evolution*, by Robert Trivers, Benjamin/Cummings, Reading, Massachusetts, 1985.

193 *"The secret of rulership":* The quotation from Orwell is in Pinker, *How the Mind Works*, p. 421.

193 *Trivers has also depicted:* "Parent-Offspring Conflict," by Robert Trivers, *American Zoologist* 14, 1974, pp. 249–264.

193 *Darwinian version of the Oedipus complex:* "Is Parent-Offspring Conflict Sex-Linked?" by Martin Daly and Margo Wilson, *Journal of Personality* 58, 1990, pp. 163–189.

194 *Buss charged:* "The Future of Evolutionary Psychology," by David Buss, *Psychological Inquiry* 6, 1995, p. 86.

194 *Many different interpretations:* See the discussion of these interpretations of quantum mechanics in Horgan, *The End of Science.*

195 *"the single best idea":* *Darwin's Dangerous Idea,* by Daniel Dennett, Simon & Schuster, New York, 1995, p. 24.

196 *Gaia, and complexity theory:* See Horgan, *The End of Science.*

196 *"The more the universe seems":* *The First Three Minutes,* by Steven Weinberg, Basic Books, New York, 1977, p. 154.

196 *Some argue that natural selection:* See "Evolution of Humans May at Last Be Faltering," by William Stevens, *New York Times,* March 14, 1995, p. C1.

196 *descendants may undergo:* See *Children of Prometheus,* by Christopher Mills, Perseus Books, Reading, Massachusetts, 1998.

196 *"There is no security against":* The quotation from Samuel Butler's 1872 novel *Erewhon* is in "Computers Near the Threshold?" by Martin Gardner, *Journal of Consciousness Studies* 3, no. 1, 1996, pp. 89–94. The essay was reprinted in Martin's book *The Night Is Large,* St. Martin's Press, New York, 1996. In the essay Gardner, the legendary science and mathematics writer, expressed his doubts about whether artificial intelligence would ever produce truly conscious, intelligent machines. When I called Gardner in January 1999 to check this citation, he told me that he considered himself a "mysterian"—someone who believes that free will, consciousness, and other aspects of the mind are mysteries that cannot be explained by science. The mysterian position is discussed in the following chapter.

197 *They have envisioned a day:* See my interviews with Marvin Minsky and Hans Moravec in *The End of Science,* Broadway Books, New York, 1996, pp. 183–188, 248–251, respectively.

197 *"Why are there so many robots":* Pinker, *How the Mind Works,* pp. 3–4.

197 *"faced the integration problem":* "The Trouble with Psychological Darwinism," by Jerry Fodor, *London Review of Books,* January 22, 1998, pp. 11–13.

PAGE **Chapter 7: ARTIFICIAL COMMON SENSE**

199 *"My expert systems":* *Galatea* 2.2, by Richard Powers, HarperCollins, New York, 1995, p. 28. Powers's novel described the effort of a novelist and cognitive scientist to build a computer that could "read" literature as well as a typical human graduate student.

199 Machines Who Think: *Machines Who Think,* by Pamela McCorduck, W. H. Freeman, San Francisco, 1979. See also McCorduck's book *The Fifth Generation,* cowritten with Edward Feigenbaum, Addison-Wesley, Reading, Massachusetts, 1983.

200 *The lead article:* "The Machine As Partner of the New Professional," by Frederick Hayes-Roth, *IEEE Spectrum,* June 1984, pp. 28–31.

201 *In 1998:* I spoke to Hayes-Roth over the telephone on January 22, 1998.

202 *Hayes-Roth is a craven defeatist:* I interviewed Herbert Simon by telephone on September 25, 1998. For more on Simon's career and views, see his books *The Sciences of the Artificial,* 3rd ed., MIT Press, Cambridge, 1996, and *Models of My Life,* Basic Books, New York, 1991.

204 *critics keep "raising the bar":* "Smart Machines, and Why We Fear Them," by Astro Teller, *New York Times,* March 21, 1998.

204 *"will help man obey":* Simon gave this speech before the annual meeting of the Operations Research Society of America in Pittsburgh, Pennsylvania, on November 14, 1957. The lecture was based on a paper cowritten with Simon's colleague Allen Newell, which was published as "Heuristic Problem Solving: The Next Advance in Operations Research," *Operations Research* 6, no. 1, January–February 1958, pp. 1–10.

205 *"not trivial and uninteresting":* McCorduck, *Machines Who Think,* p. 188.

205 *"People also see images of Jesus":* "The Artist's Angst Is All in Your Head," by George Johnson, *New York Times Week in Review,* November 16, 1997, p. 16.

205 *As for artificial mathematics:* For a discussion of the role of computers in mathematics, see my article "The Death of Proof," *Scientific American,* October 1993, pp. 92–103.

206 *"Not now, not 100 years from now":* Mumford originally made this statement to me in 1993. Mumford told me by e-mail in December 1998 that he still stood by his statement.

206 *Chess is based on:* For an excellent discussion of computer chess, see "A Grandmaster Chess Machine," by Feng-hsiung Hsu, Thomas Anatharaman, Murray Campbell, and Andreas Nowatzyk, *Scientific American,* October 1990, pp. 44–50. The authors were the creators of Deep Thought, the predecessor of Deep Blue.

207 *"This chess project is not AI":* I interviewed the Deep Blue team at IBM's Thomas J. Watson Research Center in Yorktown Heights, New York, in March 1996.

207 *"I would call what Deep Blue does":* Simon was quoted in "A Mean Chess-Playing Computer Tears at the Meaning of Thought," by Bruce Weber, *New York Times,* February 19, 1996, p. A1.

208 *IBM has discouraged its employees:* IBM's alleged aversion to an association with artificial intelligence was also reported in McCorduck, *Machines Who Think,* p. 159.

209 *such as curve interpolation:* For this background on neural networks, I am indebted to Tomaso Poggio, a professor in the Department of Brain Sciences at MIT and an authority on neural networks and machine learning. I interviewed Poggio on November 17, 1997.

209 *Hubert Dreyfus doubts:* I interviewed Hubert Dreyfus by telephone

on February 3, 1998. For a less favorable discussion of Dreyfus, see McCorduck, *Machines Who Think*, pp. 180–205.

211   *"Is an exhaustive analysis": What Computers* Still *Can't Do*, by Hubert Dreyfus, MIT Press, Cambridge, 1992, p. 303.

212   *"Computers Can't Play Chess":* McCorduck, *Machines Who Think*, p. 200.

212   *"it is now clear":* Dreyfus, *What Computers* Still *Can't Do*, p. ix.

213   *"One needs a learning device":* Ibid., p. xiv.

213   *"degenerating research program":* Ibid., p. ix.

214   *"Let's state the obvious": Hal's Legacy*, edited by David Stork, MIT Press, Cambridge, 1997, p. 5.

214   *"As we approach 2001":* Ibid., p 11.

214   *"AI so far has been a failure":* Ibid., pp. 49–50.

214   *HAL "is an unrealistic conception":* Ibid., pp. 188–189.

215   *"We're now in a position":* Ibid., pp. 201–202.

215   *"the very same brick wall":* Ibid., p. 371.

215   *"Napoleon died on Saint Helena":* Ibid., p. 203.

216   *"If you ask it what it is":* "Happy Birthday, HAL," by Simson Garfinkel, *Wired*, January 1997, p. 188.

216   *"HAL, Cyc, and their ilk":* Stork, *Hal's Legacy*, p. 207.

216   *The goal of the Cyc team:* Ibid., p. 203.

217   *"Without having to be":* Ibid., p. 206.

218   *"pick up new knowledge":* "Silicon Babies," by Paul Wallich, *Scientific American*, December 1991, p. 134.

218   *"full-fledged creative member":* Garfinkel, "Happy Birthday, HAL," p. 188. For more information on Cyc, see the Web site maintained by Lenat's company Cycorp, www.cyc.com.

218   *"I love Doug":* Brooks made these remarks when I interviewed him at MIT on November 17, 1997. See also the interview with Rodney Brooks on the Internet site the Edge, www.edge.org.

220   *Brooks wrote a series of papers:* "Fast, Cheap and Out of Control," by Brooks and Anita Flynn, *Journal of the Interplanetary System* 42, 1989; "Elephants Don't Play Chess," *Robotics and Autonomous Systems* 6, 1990; "New Approaches to Robotics," *Science*, 253, 1991; and "Intelligence Without Representation," *Artificial Intelligence* 47, 1991.

220   *Errol Morris documentary:* The philosopher David Rothenberg conducted a fascinating interview with Errol Morris for the quarterly *Terra Nova*, which Rothenberg founded and edits: "Outside the Cage Is the Cage," *Terra Nova* 3, no. 2, 1998, pp. 56–73. Brooks's work also inspired the book *Out of Control*, by Kevin Kelly, Addison-Wesley, Reading, Massachusetts, 1994.

221   *Not everyone finds Cog impressive:* The quotations from Pinker and Bever are both drawn from "Building a Baby Brain in a Robot," by John Travis, *Science*, May 20, 1994, p. 1082.

223   *the definition of a bird:* See *The Society of Mind*, by Marvin Minsky, Simon & Schuster, New York, 1985, p. 127.

223 *Minsky once told me:* I interviewed Minsky several times in person and by telephone in 1993. See also the interview with Minsky on the Edge, www.edge.org, and my discussion of Minsky in Horgan, *The End of Science*, Broadway Books, New York, 1996, pp. 183–188.

224 *"Psychoanalytic Concepts for the Control of Emotions in Robots":* Stephane Zrehen gave this lecture at a session of the fall symposium of the American Association for Artificial Intelligence, titled "Emotional and Intelligent: The Tangled Knot of Cognition." The meeting was held in Orlando, Florida, October 1998.

225 *"constitutes a decentered self":* "Artificial Intelligence and Psycho-analysis: A New Alliance," by Sherry Turkle, *Daedalus*, Winter 1988, p. 245. Turkle had previously discussed analogies between psycho-analysis and AI in her book *The Second Self*, Simon & Schuster, New York, 1984.

226 *"Trained to track": Life on the Screen*, by Sherry Turkle, Simon & Schuster, New York, 1995, p. 266.

227 *Pamela McCorduck recalled:* McCorduck, *Machines Who Think*, p. 254.

228 *innate "theory-of-mind" module:* See the discussion of the theory-of-mind theory in *How the Mind Works*, by Steven Pinker, Norton, New York, 1997, pp. 329–333.

228 *"The having of perceptions":* This quotation from Sutherland serves as an epigraph for *The Astonishing Hypothesis*, by Francis Crick, Charles Scribner's Sons, New York, 1994.

PAGE Chapter 8: THE CONSCIOUSNESS CONUNDRUM

229 *Suppose that there be:* The four quotations at the beginning of this chapter are drawn from, in order of appearance: *Consciousness Explained*, by Daniel Dennett, Little, Brown, Boston, 1991, p. 412; *How the Mind Works*, by Steven Pinker, Norton, New York, 1997, p. 132; ibid., p. 132; *The Coming of the Golden Age*, by Gunther Stent, Natural History Press, Garden City, New York, 1969, p. 74.

231 *"Toward a Scientific Basis":* The proceedings of the consciousness meeting in Tucson, which was held April 12–17, 1994, were pub-lished as *Toward a Science of Consciousness: The First Tucson Discussions and Debates*, edited by Stuart Hameroff et al., MIT Press, Cam-bridge, 1996.

232 *Steen Rasmussen, a Danish physicist:* Steen Rasmussen was profiled in "Playing God," by David Freedman, *Discover*, August 1992, pp. 35–45.

232 *veteran neuroscientist Karl Pribram:* Karl Pribram expounded on his ideas in his book *Brain and Perception*, Lawrence Erlbaum Associates, Hillsdale, New Jersey, 1991.

233 *Danah Zohar, who earned: The Quantum Self*, by Danah Zohar, William Morrow, New York, 1990.

233 *Benjamin Libet, a psychologist:* A lucid account of Benjamin Libet's re-

search can be found in *The User Illusion*, by Tor Norretranders, Viking, New York, 1998, pp. 216–220, 227–238.

235   *announced in a jointly written paper:* "Toward a Neurobiological Theory of Consciousness," by Francis Crick and Christof Koch, *Seminars in the Neurosciences* 2, 1990, pp. 263–275.

236   The Astonishing Hypothesis: *The Astonishing Hypothesis*, by Francis Crick, Charles Scribner's Sons, New York, 1994. See also my discussion of the ideas of Crick and Koch in *The End of Science*, Broadway Books, New York, 1996, pp. 159–164.

237   *Walter Freeman:* Walter Freeman presented his ideas about chaos and the brain in "The Physiology of Perception," *Scientific American*, February 1991, pp. 78–85; and in *Societies of Brains*, Lawrence Erlbaum Associates, Hillsdale, New Jersey, 1995.

238   *"Listen carefully to what":* The quotation is from Flanagan's contribution to Hameroff et al., *Toward a Science of Consciousness*. See also Flanagan's books *The Science of the Mind*, 2nd ed., MIT Press, Cambridge, 1991; and *Consciousness Reconsidered*, MIT Press, Cambridge, 1992.

239   *before Roger Penrose began:* Roger Penrose presented his ideas about quantum mechanics and the mind in *The Emperor's New Mind*, Oxford University Press, New York, 1989; *Shadows of the Mind*, Oxford University Press, New York, 1994; and *The Large, the Small and the Human Mind* (which included material from three other authors), Cambridge University Press, New York, 1997. For biting critiques of Penrose's work, see "Shadows of Doubt," by Philip Anderson, *Nature*, November 17, 1994, pp. 288–289; and "The Best of All Possible Brains," by Hilary Putnam, *New York Times Book Review*, November 20, 1994, p. 7. See also my discussion of Penrose in *The End of Science*, pp. 174–177.

242   *David Chalmers, a young Australian:* Chalmers presented his ideas in "The Puzzle of Conscious Experience," *Scientific American*, December 1995, pp. 80–87 (the article was accompanied by a sidebar by Francis Crick and Christof Koch); and in *The Conscious Mind*, Oxford University Press, New York, 1996. For a critical review of Chalmers's book, see "Consciousness and the Philosophers," by John Searle, *New York Review of Books*, March 6, 1997, pp. 43–50. John Searle is a philosopher at the University of California at Berkeley and a leading critic of the strong AI position. Searle is perhaps best known for his oddly influential "Chinese room" argument, which he laid out in "Is the Brain's Mind a Computer Program?" *Scientific American*, January 1990, pp. 26–31. (Searle's article was followed by a convoluted rebuttal written by the philosophers Paul and Patricia Churchland, "Could a Machine Think?" pp. 32–37.) Searle compared a computer in the Turing test to a man in a room who does

not understand Chinese but has a manual for converting Chinese questions or commands into Chinese responses. The man receives a string of Chinese characters that, unbeknown to him, means, "What is your favorite color?" His manual tells him that when he receives these symbols, he should respond with another string of symbols that, unbeknown to him, means "Blue." In the same way, Searle contended, computers manipulate symbols without under-standing their meaning; thus computers are not really thinking as we humans do. To my mind, Searle has not rebutted the strong AI position at all. Instead, he has merely pointed out, implicitly, how difficult it would be for a computer to pass the Turing test. A man-ual that could list all the possible questions that can be stated in Chinese, together with plausible-sounding responses for each question, would be almost infinitely long. How could the man pos-sibly respond to incoming questions fast enough to convince those outside the room that he actually understands Chinese?

246 *"the seat of the Will":* Crick, *The Astonishing Hypothesis,* p. 267.

247 *"My first act of free will":* The quotation is from "William James and the Case of the Epileptic Patient," by Louis Menand, *New York Review of Books,* December 17, 1998, p. 82.

248 *"There is no such thing":* Noam Chomsky presented me with this view of the mind-body problem during a telephone interview in 1995. He also discussed the mind-body problem, the cognitive lim-its of science, *problems* versus *mysteries,* and related ideas in *Language and Problems of Knowledge,* MIT Press, Cambridge, 1988, pp. 133–170.

249 *a position akin to vitalism:* "Facing Backwards on the Problem of Con-sciousness," by Daniel Dennett, *Journal of Consciousness Studies* 3, no. 1, pp. 4–6.

250 *Pinker concluded that consciousness:* Pinker, *How the Mind Works,* p. 561.

250 *Even the neuroscientist Christof Koch:* Christof Koch also expressed mysterian views in "Hard-Headed Dualism," *Nature,* May 19, 1996, p. 124.

251 *"It doesn't seem to me":* The philosopher Colin McGinn made these comments when I interviewed him in New York City in August 1994.

251 *"in each other's consciousness":* See Weil's contribution to Hameroff et al., *Toward a Science of Consciousness.*

252 *"a sensation of 'eternity'":* The Freud Reader, edited by Peter Gay, Nor-ton, New York, 1989, p. 723.

252 *"I cannot think of any need":* Ibid., p. 727.

252 *"none of our thoughts":* The Varieties of Religious Experience, by William James, Macmillan, New York, 1961, p. 30.

252 *"what we can ascertain":* Ibid., p. 33.

252 *"our normal waking consciousness":* Ibid., p. 305.

253   *"black-haired youth with greenish skin":* Ibid., p. 138.

253   *the awful experience had been his:* For a fascinating discussion of this episode in James's life, see Menand, "William James and the Case of the Epileptic Patient."

254   *"Physics and Spirituality":* "Physics and Spirituality: The Next Grand Unification?" by Brian Josephson, *Physics Education* 22, 1987, pp. 15–19.

254   *"function more harmoniously":* "Religion in the Genes," by Brian Josephson, *Nature,* April 15, 1993, p. 583. Other published comments by Josephson include "Skepticism and Psi: A Personal View," *Behavioral and Brain Sciences* 10, no. 4, 1987, p. 594; "Has Psychokinesis Met Science's Measure?" *Physics Today,* July 1992, p. 15; and "Consciously Avoiding the X-factor," *Physics World,* December 1996, p. 45.

254   *I had an opportunity:* I interviewed Josephson in Tucson on April 14, 1994.

256   *Bohm, in an interview:* I interviewed David Bohm in August 1992.

257   *Josephson's theory of music:* Josephson and his coauthor presented their theory of music in Hameroff et al., *Toward a Science of Consciousness.*

PAGE Epilogue: THE FUTURE OF MIND-SCIENCE

258   *"Anybody who has been seriously engaged":* The quotation is in *Bartlett's Familiar Quotations,* 15th ed., Little, Brown, Boston, 1980, p. 686. The quotation is attributed to Max Planck's 1932 book, *Where Is Science Going?*

258   *"'You,' your joys":* *The Astonishing Hypothesis,* by Francis Crick, Charles Scribner's Sons, New York, 1994, p. 3.

259   *Reductionism "is accepted":* "More Is Different," by Philip Anderson, *Science,* August 4, 1972, p. 393.

260   *"there are no large people":* *Genius,* by James Gleick, Pantheon, New York, 1992, p. 326.

261   *there may be no unifying insight:* When I was well into writing *The Undiscovered Mind,* I encountered a book that challenged my view of science in general and mind-science in particular. In *Consilience* (Alfred Knopf, New York, 1998), the evolutionary biologist Edward Wilson of Harvard University returned to a theme that he had first broached more than two decades earlier in *Sociobiology,* Harvard University Press, Cambridge, 1975. Wilson argued that it was time to resurrect the Enlightenment goal of joining all branches of knowledge—including not only science but also philosophy, history, theology, and even the arts—into a cohesive whole. Wilson defined *consilience,* a term he borrowed from the nineteenth-century philosopher William Whewell, as "literally a 'jumping together' of knowledge by the linking of facts and fact-based theory

across disciplines to create a common groundwork of explanation" (p. 8). Wilson's proposal raised two questions: First, is it feasible? If science is to be joined to the arts and humanities, the juncture will surely be through those fields that address the human mind. Recognizing this fact, Wilson surveyed behavioral genetics, evolutionary psychology, artificial intelligence, and neuroscience. Although our knowledge of our own minds remains fragmentary, he concluded, the pieces would one day be pulled together into a coherent theory. "The grand synthesis could come quickly, or it could come with painful slowness over a period of decades" (p. 109).

Note that decades is the *pessimistic* estimate. But given the rancorous disagreements between endeavors as closely related as sociobiology and evolutionary psychology, what hope is there for consilience between, say, particle physics and literary criticism? Accepting, for a moment, that the consilience Wilson envisioned is possible, one still faces the question of whether it is desirable. In other words, what's the upside? Wilson did not shy away from the question. "It is worth asking, particularly in the present winter of our cultural discontent, whether the original spirit of the Enlightenment—confidence, optimism, eyes to the horizon—can be regained. And to ask in honest opposition, *should* it be regained, or did it possess in its first conception, as some have suggested, a dark angelic flaw? Might its idealism have contributed to the Terror, which foreshadowed the horrendous dream of the totalitarian state?" (p. 21). Wilson never really answered his own questions. His great hope seemed to be that self-understanding will compel us to accept what for Wilson is a particularly urgent truth: "To the extent that we depend on prosthetic devices to keep ourselves and our biosphere alive, we will render everything fragile. To the extent that we banish the rest of life, we will impoverish our species for all time" (p. 298). It is hard to see how this worthy goal—the preservation of nature—would be served by the discovery and acceptance of a unified theory of knowledge based on evolutionary biology, genetics, and neuroscience.

Many people who disagree with Wilson on the feasibility and desirability of such a theory can and do embrace his brand of environmentalism. Conversely, many who accept his reductionist vision of humanity no doubt find his conservationist ethic too extreme. In fact, Wilson may actually damage the cause of conservation by linking it so vehemently to his reductionist and, yes, deterministic view of human nature. Ironically, when I spoke to Wilson in 1994, he seemed to believe that a final theory of human nature might be neither desirable nor feasible. (See Horgan, *The*

End of Science, pp. 143–149.) He feared that such a theory would reduce "our exalted self-image, and our hope for indefinite growth in the future"; it might also spell the end of biology, the discipline that had given meaning to his own life. Wilson had resolved this dilemma by deciding that the mind could never be completely understood; the interaction between nature and nurture, genes and culture, represented "an immense unmapped area of science and human history that we would take forever to explore." Not decades, but forever.

262 *The attempt to harness nuclear fusion:* The pros and cons of fusion energy were fiercely debated in the "Letters" section of *Physics Today* in March and May 1997. The exchange concluded in the May issue with a letter cowritten by three authorities on fusion: William Parkins, James Krumhansl, and Chauncey Starr. They stated: "In the case of fission, a remarkably fortuitous set of technical properties made today's nuclear power industry possible. In the case of fusion, a very unfortunate set of constraints appears to obviate any future power industry based on the fusion principle."

263 *cancer mortality rates:* See "Cancer Undefeated," by John Bailar and Heather Gornik, *New England Journal of Medicine*, May 29, 1997, pp. 1569–1574. The article reported that age-adjusted cancer mortality rates in the United States have risen by 6 percent since 1970 and by more than 8 percent since 1950. When the rates are not adjusted for the aging of the population, the increases are much more alarming.

263 *"In the great majority of cases":* Great and Desperate Cures, by Elliot Valenstein, Basic Books, New York, 1986, p. 249.

263 *Isaiah Berlin warned:* The quotations from Isaiah Berlin are in *Why Freud Was Wrong*, by Richard Webster, Basic Books, New York, 1995, pp. 444–445.

264 *Sacks once told me:* I briefly spoke to Sacks by telephone in the fall of 1997. In spite of his antireductionist sympathies, Sacks has often expressed admiration for a controversial theory of mentation promulgated by Gerald Edelman, a Nobel laureate and director of the Neurosciences Institute in La Jolla, California. See my discussion of Edelman in *The End of Science*, Broadway Books, New York, 1996, pp. 165–172.

264 *"To restore the human subject":* The Man Who Mistook His Wife for a Hat, by Oliver Sacks, Harper Perennial, New York, 1987, p. viii.

265 *"The realities of patients":* An Anthropologist on Mars, by Oliver Sacks, Vintage Books, New York, 1996, pp. xviii–xix.

266 *such experiences "cannot be ignored":* The Doors of Perception and Heaven and Hell, by Aldous Huxley, Harper & Row, New York, 1990, p. 84. Huxley wrote the two essays that comprise the book in the 1950s.

Notes

266   *"Not* how *the world is":* Tractatus Logico-Philosophicus, by Ludwig
       Wittgenstein, Routledge, New York, 1990, p. 187.
267   *Lilly described the Beings:* For a glimpse of John Lilly's worldview, see
       his two autobiographies: *The Center of the Cyclone,* Bantam Books,
       New York, 1973, and *The Scientist,* Ronin Publishing, Berkeley, Cali-
       fornia, 1988.

SELECTED
BIBLIOGRAPHY

Barkow, Jerome, Cosmides, Leda, and Tooby, John, editors, *The Adapted Mind*, Oxford University Press, New York, 1992.

Breggin, Peter, *Toxic Psychiatry*, St. Martin's Press, New York, 1991.

Breggin, Peter, and Breggin, Ginger, *Talking Back to Prozac*, St. Martin's Press, New York, 1994.

———, *The War Against Children*, St. Martin's Press, New York, 1994.

Buss, David, *The Evolution of Desire*, Basic Books, New York, 1994.

Chalmers, David, *The Conscious Mind*, Oxford University Press, New York, 1996.

Chomsky, Noam, *Language and Problems of Knowledge*, MIT Press, Cambridge, 1988.

Crenshaw, Teresa, and Goldberg, James, *Sexual Pharmacology: Drugs That Affect Sexual Function*, Norton, New York, 1996.

Crews, Frederick, *The Memory Wars*, New York Review of Books, New York, 1995.

———, editor, *Unauthorized Freud*, Viking, New York, 1998.

Crick, Francis, *The Astonishing Hypothesis*, Charles Scribner's Sons, New York, 1994.

Damasio, Antonio, *Descartes's Error*, Avon Books, New York, 1994.

Dawes, Robyn, *House of Cards*, Free Press, New York, 1994.

Degler, Carl, *In Search of Human Nature*, Oxford University Press, New York, 1991.

Dennett, Daniel, *Consciousness Explained*, Little, Brown, Boston, 1991.

Dolnick, Edward, *Madness on the Couch*, Simon & Schuster, New York, 1998.

Dreyfus, Hubert, *What Computers Still Can't Do*, MIT Press, Cambridge, 1992.

Fisher, Seymour, and Greenberg, Roger, *Freud Scientifically Reappraised*, John Wiley & Sons, New York, 1996.

———, editors, *From Placebo to Panacea*, John Wiley & Sons, New York, 1997.

Fishman, Daniel, *The Case for Pragmatic Psychology*, New York University Press, New York, 1999.

Flanagan, Owen, *The Science of the Mind*, 2nd ed., MIT Press, Cambridge, 1991.

————, *Consciousness Reconsidered*, MIT Press, Cambridge, 1992.

Frank, Jerome, and Frank, Julia, *Persuasion and Healing*, 3rd ed., Johns Hopkins University Press, Baltimore, 1993.

Freeman, Walter, *Societies of Brains*, Lawrence Erlbaum Associates, Hillsdale, New Jersey, 1995.

Gardner, Howard, *Frames of Mind*, Basic Books, New York, 1983.

————, *The Mind's New Science*, Basic Books, New York, 1985.

————, *Extraordinary Minds*, Basic Books, New York, 1997.

Gardner, Howard, Kornhaber, Mindy, and Wake, Warren, *Intelligence: Multiple Perspectives*, Harcourt Brace, New York, 1996.

Gay, Peter, editor, *The Freud Reader*, Norton, New York, 1989.

Geertz, Clifford, *Works and Lives*, Stanford University Press, 1988.

Gross, Charles, *Brain, Vision, Memory: Tales in the History of Neuroscience*, MIT Press, Cambridge, 1998.

Hamer, Dean, and Copeland, Peter, *The Science of Desire*, Simon & Schuster, New York, 1994.

————, *Living with Our Genes*, Doubleday, New York, 1998.

Hameroff, Stuart, et al., editors, *Toward a Science of Consciousness: The First Tucson Discussions and Debates*, MIT Press, Cambridge, 1996.

Harrington, Anne, editor, *The Placebo Effect*, Harvard University Press, Cambridge, 1997.

Harris, Judith, *The Nurture Assumption*, Free Press, New York, 1998.

Hooper, Judith, and Teresi, Dick, *The Three-Pound Universe*, St. Martin's Press, New York, 1986.

Horgan, John, *The End of Science*, Broadway Books, New York, 1996.

Huxley, Aldous, *The Doors of Perception and Heaven and Hell*, Harper & Row, New York, 1990.

James, William, *The Varieties of Religious Experience*, Macmillan, New York, 1961.

Jamison, Kay, *An Unquiet Mind*, Vintage Books, New York, 1995.

Johnson, George, *In the Palaces of Memory*, Vintage Books, New York, 1992.

Kagan, Jerome, *Galen's Prophecy*, Basic Books, New York, 1994.

————, *Three Seductive Ideas*, Harvard University Press, Cambridge, 1999.

Kandel, Eric, Schwartz, James, and Jessel, Thomas, editors, *Essentials of Neural Science and Behavior*, Appleton & Lange, Stamford, Connecticut, 1995.

Kevles, Daniel, *In the Name of Eugenics*, Alfred A. Knopf, New York, 1985.

Kramer, Peter, *Listening to Prozac*, Penguin Books, New York, 1993.

Kutchins, Herb, and Kirk, Stuart, *Making Us Crazy*, Free Press, New York, 1998.

LeDoux, Joseph, *The Emotional Brain*, Simon & Schuster, New York, 1996.

Macmillan, Malcolm, *Freud Evaluated*, MIT Press, Cambridge, 1997.

Malcolm, Janet, *The Impossible Profession*, Vintage Books, New York, 1982.

Mayr, Ernst, *Toward a New Philosophy of Biology*, Harvard University Press, Cambridge, 1988.

McCorduck, Pamela, *Machines Who Think*, W. H. Freeman, San Francisco, 1979.

McCorduck, Pamela, and Feigenbaum, Edward, *The Fifth Generation*, Addison-Wesley, Reading, Massachusetts, 1983.

Mills, Christopher, *Children of Prometheus*, Perseus Books, Reading, Massachusetts, 1998.

Minsky, Marvin, *The Society of Mind*, Simon & Schuster, New York, 1985.

Mithen, Steven, *The Prehistory of the Mind*, Thames and Hudson, London, 1996.

Nesse, Randolph, and Williams, George, *Why We Get Sick*, Times Books, New York, 1994.

Norretranders, Tor, *The User Illusion*, Viking, New York, 1998.

Penrose, Roger, *The Emperor's New Mind*, Oxford University Press, New York, 1989.

————, *Shadows of the Mind*, Oxford University Press, New York, 1994.

Pinker, Steven, *The Language Instinct*, Harper Perennial, 1994.

————, *How the Mind Works*, Norton, New York, 1997.

Pressman, Jack, *Last Resort*, Cambridge University Press, New York, 1998.

Roth, Michael, editor, *Freud: Conflict and Culture*, Knopf, New York, 1998.

Sacks, Oliver, *The Man Who Mistook His Wife for a Hat*, Harper Perennial, New York, 1987.

————, *An Anthropologist on Mars*, Vintage Books, New York, 1995.

Schacter, Daniel, *Searching for Memory*, Basic Books, New York, 1996.

Shorter, Edward, *A History of Psychiatry*, John Wiley & Sons, New York, 1997.

Silver, Lee, *Remaking Eden*, Avon Books, New York, 1997.

Simon, Herbert, *The Sciences of the Artificial*, 3rd ed., MIT Press, Cambridge, 1996.

Stent, Gunther, *The Coming of the Golden Age*, Natural History Press, Garden City, New York, 1969.

————, *Paradoxes of Progress*, W. H. Freeman, San Francisco, 1978.

Stork, David, editor, *Hal's Legacy*, MIT Press, Cambridge, 1997.

Sulloway, Frank, *Freud, Biologist of the Mind*, Basic Books, New York, 1979.

————, *Born to Rebel*, Pantheon Books, New York, 1996.

Torrey, E. Fuller, *Freudian Fraud*, HarperCollins, New York, 1992.

Turkle, Sherry, *The Second Self*, Simon & Schuster, New York, 1984.

————, *Life on the Screen*, Simon & Schuster, New York, 1995.

Valenstein, Elliot, *Great and Desperate Cures*, Basic Books, New York, 1986.

Vaughan, Susan, *The Talking Cure*, G. P. Putnam's Sons, New York, 1997.

Webster, Richard, *Why Freud Was Wrong*, Basic Books, New York, 1995.

Wilson, Edward, *Sociobiology*, Harvard University Press, Cambridge, 1975.

————, *On Human Nature*, Harvard University Press, Cambridge, 1978.

————, *Consilience*, Alfred Knopf, New York, 1998.

Wolpert, Lewis, *The Unnatural Nature of Science*, Harvard University Press, Cambridge, 1993.

Wright, Lawrence, *Twins*, John Wiley & Sons, New York, 1997.

Wright, Robert, *The Moral Animal*, Pantheon Books, New York, 1994.

ACKNOWLEDGMENTS

I am indebted to those who critiqued part or all of this book for no remuneration. They include Roger Bingham, Chris Bremser, Walter Brown, Robyn Dawes, Hubert Dreyfus, Roger Greenberg, Fred Guterl, Judith Harris, Jerome Kagan, Christof Koch, Mindy Kornhaber, Eric Kramer, Robert Plomin, Phil Ross, David Rothenberg, Ellen Shell, Gary Stix, Karen Wright, and Robert Wright. Of course, none of them necessarily endorses the book's contents, and all errors are my responsibility alone. I am grateful to my editor, Stephen Morrow, and my agent, John Brockman, for their personal and professional support. And thanks above all to Suzie, who reads me all too well.

INDEX

homelessness, 95, 100, 140
Homicide (Wilson and Daly), 184, 295–296
homosexuality, 6, 56, 58, 79, 140, 141, 152–154, 187, 290
Hooper, Judith, 282
hopeful skepticism, 13–14, 264
House of Cards (Dawes), 94–95, 281
Houston, Whitney, 257
How the Mind Works (Pinker), 50, 168, 175–176, 186, 192, 197, 250, 275, 294, 295, 297–298, 301, 303
How We Die (Nuland), 4, 269
Hrdy, Sarah Blaffer, 191, 295
Hubel, David, 19–20, 22–23, 27, 271
Hudson River Psychiatric Center, 96–100, 281
Huffington, Ariana, 283
Human Behavior and Evolution Society (HBES), 167–172, 174, 179–180, 182, 188, 190, 292, 294–296
Human Genome Project, 140
Humpty Dumpty dilemma, 23, 43, 198, 236, 262
hunter-gatherers, 168–169, 174, 180. *See also* tribal societies
Huntington's disease, 138, 147, 161
Huxley, Aldous, 117, 266, 306
Huxley, Thomas, 229
hydrotherapy, 106, 253
Hyman, Steven, 67–69, 276
hysteria, 49, 59, 79

IBM, 82, 200, 205, 207–208, 240, 299
Identity: Youth and Crisis (Erikson), 75
Illiac Suite, 205
I'm Okay—You're Okay (Harris), 75
imipramine, 116, 118–119. *See also* tricyclics
In Search of Human Nature (Degler), 292
In the Name of Eugenics (Kevles), 287, 291
incest, 55, 64, 65, 110
infanticide, 183–186, 295–296
information, 242–243
information theory, 223, 233
insulin-coma therapy, 106
insurance, 50, 77, 80, 94, 103, 116, 120. *See also* managed health care

intelligence, 5, 25, 26, 182, 196, 203, 211, 212, 216, 220, 243, 245, 291–292
genes and, 84, 137, 140, 144, 154–164
multiple, 23, 71, 155
race and, 13, 84, 155–159
See also artificial intelligence; common sense; IQ
International Dictionary of Psychology (Sutherland), 228
International Psychoanalytic Association, 51
interpersonal therapy, 118–119, 277
Interpretation of Dreams, The (Freud), 49, 54
ironic science, 6
Isaacson, L.M., 205
IQ, 6, 71, 72, 139, 146, 154–161, 291
race and, 84, 155–159
See also intelligence

James, William, 9, 72–73, 113, 247, 252–253, 303–304
Jamison, Kay, 122–123, 285
Janov, Arthur, 75
Jaroff, Leon, 287
Jensen, Arthur, 155, 158, 291
Jesus, 5, 205
Johnny Carson Show, 146
Johnson & Johnson, 124
Johnson, George, 271, 273, 299
Johnson, Samuel, 229
Josephson, Brian, 254–257, 304
Journal of the American Medical Association, 116, 151–152
Journal of the American Psychoanalytic Association, 82
Judd, Lewis, 37, 272
Jung, Carl, 75, 76
Jungian therapy, 6, 78

Kagan, Jerome, 37, 162–164, 272, 292
Kamin, Leon, 145, 154, 288
Kandel, Eric, 40–44, 54, 66, 129, 246, 260, 273
Karasu, Toksoz, 78–79, 277, 278, 281
Kasparov, Gary, 207, 212, 240
Kelly, Kevin, 300
Kesey, Ken, 108
ketamine, 267

ABOUT THE AUTHOR

John Horgan is a freelance writer and author of *The End of Science*, a U.S. best-seller that has been translated into ten languages. His awards include the Science Journalism Award of the American Association for the Advancement of Science (1992 and 1994) and the National Association of Science Writers Science-in-Society Award (1993). He has written for the *New York Times, London Times, Washington Post, New Republic, Slate, Discover, The Sciences,* and other publications in the United States and Europe. He was a staff writer at *Scientific American* from 1986 to 1997 and at *IEEE Spectrum* from 1983 to 1986. He graduated from Columbia University's School of Journalism in 1983. He lives in Garrison, New York, with his wife, Suzie Gilbert, a childrens' book author, and their two children.